COMPETITION FOR SCARCE GROUNDWATER IN THE SANA'A PLAIN, YEMEN

*A STUDY ON THE INCENTIVE SYSTEMS FOR URBAN AND AGRICULTURAL WATER USE*

# Competition for Scarce Groundwater in the Sana'a Plain, Yemen

## *A study on the incentive systems for urban and agricultural water use*

DISSERTATION
Submitted in fulfilment of the requirements of
the Board for Doctorates of Delft University of Technology
and the Academic Board of the International Institute for Infrastructural,
Hydraulic and Environmental Engineering for the Degree of DOCTOR
to be defended in public
on Tuesday, 7 November 2000 at 13:30 h
in Delft, the Netherlands

*by*

MOHAMED I. AL-HAMDI
*B.Sc. in Civil Engineering, George Washington University, Washington DC, USA*
*M.Sc. in Sanitary Engineering, IHE, Delft, Netherlands*
*born in Sana'a, Republic of Yemen*

A.A. BALKEMA / ROTTERDAM / BROOKFIELD / 2000

This dissertation has been approved by the promoters:

| | |
|---|---|
| Prof. dr ir G.J. Alaerts | International Institute for Infrastructural, Hydraulic and Environmental Engineering/Delft University of Technology |
| Prof. dr W.G. Wolters | Katholieke Universiteit Nijmegen |

Other committee members:

| | |
|---|---|
| Chairman | Rector Magnificus, Delft University of Technology |
| Co-chairman | Rector, International Institute for Infrastructural, Hydraulic and Environmental Engineering |
| Prof. dr ir C. van den Akker | Delft University of Technology |
| Prof. dr M.L. Al-Eryani | Sana'a University, Yemen |
| Prof. dr ir H.H.G. Savenije | International Institute for Infrastructural, Hydraulic and Environmental Engineering |
| Mr C. Ward | Principal Economist, The World Bank, Washington DC, USA |



Published by
A.A. Balkema, P.O. Box 1675, 3000 BR Rotterdam, Netherlands
Fax: +31.10.4135947; E-mail: balkema@balkema.nl; Internet site: http://www.balkema.nl

A.A. Balkema Publishers, 2252 Ridge Road, Brookfield, VT 05036-9704, USA
Fax: 802.276.3837; E-mail: info@ashgate.com

ISBN 90 5410 426 0


Printed in the Netherlands

*To my father*

# Table of Content

## Summary

The Sana'a basin is located in the central highland of Yemen covering an area of around 3,200 km$^2$. The population in the Sana'a basin relies to a large extent on groundwater for both municipal as well as for irrigated agriculture. Groundwater has been heavily exploited since the early 1970s, where in 1990, the total groundwater abstraction was estimated at 180 Mm$^3$, of which 83% is thought to have been withdrawn for irrigation. As of 1995, the extensive groundwater withdrawal has resulted in an annual deficit of around 120 Mm$^3$ or 300% of the total natural recharge, resulting in an annual 3-6 m drop in water levels.

This study does not intend to devise detailed operational tools for demand management to control the current state of groundwater mining. Rather, it aims at identifying the main incentive frameworks that influence decision making by the key water users, namely the urban dwellers and the farmers in the Sana'a basin.

**Chapter 1**
With undefined property rights and non-exclusive access, overexploitation of groundwater in water scare regions is a typical Common-pool-resources (CPR) problem. The competition over a CPR leads users to attribute less value to benefits that they expect to receive in the future and more value to present gains. Under such conditions the "rule of capture" usually governs the use of the resource stocks and may lead to complete destruction of the resource. Without a fair, orderly and efficient method to allocate and distribute the resource units, local appropriators have little incentive to contribute to sustainable use of the resource system.

To control competition over the resource and thus raise the value of the resource units, property rights need to be clearly defined. This usually takes place once boundary rules, to identify the conditions for access the resource, and authority rules, to specify the allocation structure of the resource units to the different appropriators, are identified and enforced. For sustainable use of a resource, authority rules need to be devised in accordance with the safe yield of the resource. Having devised boundary and authority rules, commitment by the users to follow those rules is not automatic. In the presence of substantial private benefits, the temptation to break the rules is high, especially given the uncertainty of compliance by others. To solve the commitment problem, users need to accept neutral monitoring and be willing to sanction violators. For effective monitoring, reliable information, not only on the physical attributes of the resource (boundary, yield, etc.), but also information related to the use of the resource and the level of compliance, are to be supplied and maintained.

**Chapter 2** describes and analyzes the structure and cost of the domestic water supply market in Sana'a. The current annual domestic water demand of 26 Mm$^3$ is expected to increase to 45 and 95 Mm$^3$/yr for the year 2010 and 2025, respectively. The type and level of water supply and consequently the price and share of water in the household budget in Sana'a seems to be governed by location of the household. With large price variation between the different supply alternatives, poor households relying on private tanker supply for non-potable water and filtered drinking water from supermarkets are the most effected. The total share of water (potable, non-potable and sanitation services) in the budget of a typical household ranges from 1.5-

13.5%, depending on the type of service and income. With low cost recovery potential and the rapid depletion of the local groundwater resources, the public water supply is constrained to maintain its coverage and invest in new expensive supply schemes. Water transfer from other regions (Marib, Surdud, or desalinized water from the coastal area) is expected to face major social, political, financial and technical constraints.

**Chapter 3** analyzes the consequences of direct disposal of raw domestic sewage on the quality of water supply (groundwater and the distribution system) in Sana'a. Over half of the city's population relies on unregulated private water vendors, who in turn fetch water from unmonitored private wells scattered within the city's perimeter. From the health and environmental viewpoints this is alarming since around 75% of the population, or over 700,000 inhabitants, discharge their domestic wastewater to on-site cesspits, thus threatening to contaminate the same aquifers used for private domestic water supply. The pattern of groundwater contamination correlates reasonably well with population density, as groundwater under the central zone of the city tends to be more contaminated than that under the less densely populated southern and northern zones. This contamination threat has lead a large portion of the population to rely on private unregulated small filtration stations for drinking water. The large fluctuations in water pressure in the public water distribution network (intermittent supply) in the Nokom quarter of Sana'a induces back-pressure and suction of wastewater from the sub-soil into the network, thus contaminating water supply in the network.

**Chapter 4** examines the legal and cultural status of groundwater in the Sana'a basin to clarify the issues of groundwater ownership, use rights and the resolution mechanisms of water related conflict. In the absence of a water law in Yemen, groundwater management has been largely left to the users (domestic (both public and private), industry, and agriculture. The absence of a detailed legal framework, the lack of proper institutional setup to administer and manage water resources, and the weak capacity of the government to implement and enforce water related policies have allowed the local users to mine groundwater. While the constitution declares all natural resources to be state property, the *shari'ah* considers groundwater to be a communal property, but grants the well owner private ownership (tradable and salable) of all the water that percolates to his well. Most farmers in the Sana'a basin regard groundwater to be the property of the owner of the overlaying land, although in reality, access to groundwater is governed by the share in a well regardless of the farm size. The tribe is displayed as a coherent socio-political institution with consistent framework for water allocation and for conflict resolution, and identify the sheik (both at the village and tribe level) as a key figure in decision-making and conflict resolution.

**Chapter 5** examines the agronomic aspects of groundwater use in the cultivation of qat and grapes. The heavy groundwater development that took place during the 1980s, in combination with extensive sharing of boreholes among farmers, has provided a large portion of the basin population with a reliable water source, thus encouraging rapid expansion of irrigated agriculture. Extensive well sharing enhances access to a reliable irrigation source, thus increasing equity, and productivity; the shift to more valuable crops increased also the value of groundwater to the farmers. However, this has only been possible at the expense of the resource sustainability. Farmers in the basin area behave rationally, but are strongly risk-avert in crop choice, agronomy and irrigation habits. While grapes were given priority in irrigation scheduling, qat was given more attention in the quantity of irrigation water often resulting in over-irrigation. Water availability has changed the cropping pattern in the Sana'a basin area, converting large areas of traditional rain harvesting land into qat cultivation, and uncultivated land and land under cereals into grape production.

**Chapter 6** evaluates the cost and profit generating value of groundwater use for qat and grape cultivation in the Sana'a basin, and examines the effects of higher input costs on groundwater use for qat and grape cultivation. Excluding the capital cost of land and initial plantation, the share of water related costs in the overall cultivation costs of qat and grapes was in the order of 60-70%. The average total annual cultivation costs were around 2,000 and 1,500 US$/ha for qat and grapes, respectively. The 67% rise in the price of diesel at the end of 1997 resulted in an increase of the fuel cost component in the water-related costs from 28% in 1997 to 35% in 1998. The share of fuel (diesel) in the overall cultivation costs, the water related costs, and the O&M costs was 12-26%, 19-34%, and 29-56%, respectively. Land and water profitability was in the range of 8,000-13,000 US$/ha and 0.5-0.8 US$/m$^3$ for qat and 3,000-7,000 US$/ha and 0.4-0.8 US$/m$^3$ for grapes. The cost-profit ratio of qat and grape cultivation is in the range of 15-25% and 25-45%, respectively. Given the high return from irrigation, financial incentives targeted at the farmer's cost are likely to fail, and to address the water scarcity in the Sana'a basin, priorities in water allocation should be made in a more participatory, explicit and transparent manner.

**Chapter 7** analyzes the reuse possibilities of domestic wastewater in Sana'a, including a review of the current local wastewater reuse practice, an assessment of the available wastewater in terms of both quantity and quality, and identifying and valuing six reuse scenarios on the basis of five different criteria. As of 1997, the collected portion of wastewater totaled around 5.2 Mm$^3$ per year, which translates into a sewerage coverage of around 20-25%. Collected wastewater is being utilized by farmers to irrigate fodder and cereal crops along the effluent channel. At present, 75-80% of the total raw sewage from the city is discharged to the underground via cesspits, thus artificially recharging the local aquifers, but also causing unchecked local groundwater contamination. Even though direct irrigation of fodder and cereal crops is considered financially beneficial to a small number of farmers, it has not contributed to a new reduction in groundwater withdrawals. At present, the quality of the wastewater makes it unsuitable for agricultural reuse, which is mainly attributed to the unsatisfactory treatment in the overloaded oxidation ponds and the high percentage of untreated bypassed flow. With the new treatment plant (under construction), there will be greater capacity to withstand future expansion of the sewerage system and produce adequate quality effluent for further reuse. Given the local circumstances of the Sana'a basin, evaluation of six reuse alternatives shows that groundwater recharge and recovery to partially replace groundwater irrigation of cash crops irrigation is likely the most appropriate alternative.

**Chapter 8** synthesizes the incentive system that promote unsustainable use of groundwater in the Sana'a basin. Factors (disincentives) that hinder the stakeholders (farmers) in the Sana'a basin to initiate actions to establish self-organized institutions to deal with over-exploitation of groundwater are: (1) the large size of the basin area increasing the complexity of the problem and introduces higher uncertainties and greater heterogeneity among the pumpers, (2) lack of accurate information, where the available information has not been widely distributed among the users, (3) high discount rates with groundwater being critical to the users' main economic activities and welfare, and (4) the inability of the existing political regime to provide the facilities and support to the local pumpers to initiate institutional change. As for the city's dwellers, it seem that although they pay much attention to the day-to-day water problem on its "micro" scale (e.g., where to obtain the next tanker load) they do not to be concerned with the overall gloomy "macro" scale of the problem. It is believed that the confusing and unreliable information and the lack of awareness reflect a perception that the water problem, at least on its large-scale, is the government's problem to solve.

# Chapter 1

# Introduction: The Supply of and Demand for Water Resources in the Sana'a Basin

## 1.1 Water availability and use in Yemen

### 1.1.1 Socio-economic background

The Republic of Yemen (ROY) was established on May 22, 1992, when the Yemen Arab Republic (North Yemen) and the People's Democratic Republic of Yemen (South Yemen) were reunified. The ROY is located at the south-southwest part of the Arabian Peninsula between 12° and 19° north of the equator and between 42° and 55° east of Greenwich. The topography of the country varies widely between sea level in the western and southern coastal plains to elevations of more than 3,700 meters above sea level (masl) in the northwest mountains of the country. Yemen's capital, Sana'a, is situated in a large plain in the central highlands at 2,200 m altitude. Yemen covers a total area of around 536,000 $(km)^2$ and consists of 19 governorates. The latest national population census (December 1994) showed a total population of 15,831,757 inhabitants, of whom 23.24% live in urban areas, and 0.97 million in Sana'a. The annual national population growth rate was reported at 3.7%, with an elevated rate of 8.9% in the urban areas, which is mostly attributed to internal migration from the rural areas. The population of Yemen has been projected to reach 20.3 and 23.4 million by the year 2005 and 2010, respectively (Central Bureau of Statistics, 1995; Van der Gun and Ahmed, 1995).

In 1990 the per capita Gross Domestic Product (GDP) was estimated at 760 US$. Governmental services and agriculture dominate the economy of the country, with 21.4% and 20.6% shares in the GDP in 1990, respectively. Although agriculture contributed less to the GDP than the service sector, it employed around 60% of the national labor force (Van der Gun and Ahmed, 1995). In 1998, according to the World Bank (1999), the share of agriculture, industry, and services in the GDP in 1998 was 16.2%, 44.9% and 38.9% respectively. The GDP was estimated at 5.4 US$ billion, lowering the per capita GDP to around 300 US$, which is very low compared to the average of 2,025 US$ in the Middle East and North Africa region.

Poor life expectancy in Yemen of 54 years is attributed to inadequate health service and poor water and sanitation coverage. It is low when compared with the average of 67 years in the Middle East and North Africa region. Poverty in Yemen is high, with 19[1]-26.9[2]% of the households living below the poverty line (Ministry of Planning and Development, 1998; World Bank, 1999).

[1] According to World Bank (1999).
[2] According to Ministry of Planning and Development (1998).

### 1.1.2 Water supply and sanitation

Coverage of the public water supply in 1990 was estimated to be 61% of the population in the urban areas and 47% in the rural areas. With exception of Aden and Dhamar, all cities rely on intermittent supply of water, where for example water is supplied to connected consumers every 2-4 days in Sana'a and every 1-2 weeks in Taiz. The cause of intermittent supply can be capacity problems in the infrastructure (Sana'a) or lack of water resources (Taiz). Overall access to safe water has been estimated to be 39%. With regard to sanitation, 10.6% of the national population has been estimated to have access to adequate sanitation services covering 40% and 1.2% of the urban and rural population, respectively. The higher access to sanitation services in the urban areas is due to the construction and installation of partial sewerage systems in the cities of Sana'a, Aden, Mukalla, Taiz, Hodeidah, Ibb, and Dhamar, collecting around 33 $Mm^3$/a of sewage in 1990 (HWC, 1992a; Ministry of Planning and Development, 1998). The rest of the urban population relies mostly on on-site facilities (cesspits) for wastewater disposal. Unregulated on-site disposal of wastewater is of a major health and environmental concern, especially in urban, densely populated areas, placing the general public at risk from direct exposure to sewage in areas where the soil has reached its absorption capacity and sewage leaks onto the streets, and threatening to contaminate the groundwater aquifers used for domestic supplies.

Wastewater treatment plants (oxidation pond systems) have been constructed in Aden, Taiz, Hodeidah, and Dhamar. In Ibb however, an activated sludge plant was constructed in the mid 1980s. In Sana'a, the collected sewage is diverted to temporary overloaded oxidation ponds, which are to be shortly replaced by the newly constructed activated sludge plant. Effluent from the treatment plants of the various cities is mostly evaporating and leaching to the soil, with the remainder used for agriculture to irrigate mostly cereals and fodder crops. Given the extreme scarcity of water in Yemen, wastewater should not be wasted, but rather assessed and utilized in the overall context of water resources management.

### 1.1.3 Water resources

The total renewable water in Yemen is estimated at 2.1 $Gm^3$/a, which translates into a share of only 130 $m^3$/cap./a (World Bank, 1997). This is rather low when compared with the Middle East and North Africa average of 1,250 $m^3$/cap./a and a world average of 7,500 $m^3$/cap./a (Berkoff, 1994). With the relatively high population growth, the per capita share of renewable water in Yemen is expected to further decline to 90 and 72 $m^3$/cap./a by the year 2010 and 2025, respectively. These figures reflect a very high level of water scarcity in Yemen, especially when compared with the stress level of 1,700 $m^3$/cap./a considered as the minimum requirement for human needs (food production and domestic use) (Postel, 1992). As of 1994, the total water use in Yemen was estimated to have been around 2.8 $Gm^3$/a, leading to a deficit of 0.7 $Gm^3$/a. This deficit is mostly being overcome through the use of around 45,000 private wells withdrawing fossil groundwater, leading to groundwater mining and resulting in a decline of water levels in most aquifers. Agriculture is by far the largest water-using sector in Yemen, followed by the municipal and industrial sectors with shares of 93.1%, 4.6% and 2.3%, respectively. It is further estimated that 34% of the total agricultural land (1,052,000 hectare) is being irrigated with groundwater (World Bank, 1997). Prior to the 1970s, Yemen was very secluded from the rest of the world, and nearly all agriculture was rain-fed. This, combined with low population pressure, in effect meant that the country's water use was sustainable.

Within the northern, mountainous governorates, groundwater abstraction in 1990 was estimated at 1.5 $Gm^3/a$, while the recharge to the same aquifers was estimated at around 1.0 $Gm^3/a$, indicating that abstraction had exceeded recharge by around 50% (HWC, 1992b). Given the estimated usable storage of those aquifers (to a depth of 150 m) to be around 35 $Gm^3$, and assuming a constant level of abstraction (which is an optimistic view), the total usable storage would be depleted within less than 70 years. Although groundwater mining is a common feature for the entire country, heavy mining is posing a serious problem in the northern high plain region, and particularly the Sana'a basin with an annual decline of 2-6 m in groundwater levels. Groundwater use has been accelerated during the last two decades due to the absence of any administrative or traditional control over drilling.

According to the classification proposed by UNESCO, the climate of Yemen varies from hyper-arid (in the deserts, most of the plateau, and parts of the coastal plains) to sub-humid (scattered wetter zones in the western and southern slopes) with a few humid spots on a very small scale. Despite shortages in extensive and reliable data, precipitation in general is considered to vary between 50 mm/a in the coastal plains to a range of 300-700 mm/a in the central and northern high plains. Average rainfall values higher than 250 mm/a are only observed in the western and southern parts of the mountain massive. Low rainfall is found particularly in Almaharah governorate and the northern part of Hadramawt governorate. Most regions of the country have two distinct rainfall patterns, with a first rainy season in spring (March-May) and a second during the summer (July-September) (Van der Gun and Ahmed, 1995). Rainfall usually comes in short intensive events or squalls. This often leads to temporary flooding that is evacuated via wadis. With the exception of small permanent springs in the few water-rich regions, e.g., Ibb, the country does not have perennial rivers.

The introduction of drilling rigs and powerful pumps in the late 1960s has enabled the abstraction of large quantities of groundwater from deep boreholes. Farmers have been active in groundwater development during the last two decades. Water allocation generally follows the traditional principles of the Islamic system. Traditional spate irrigation is governed by the principle of *Ala'ala-fa-ala'ala,* giving upstream land senior irrigation rights over downstream land. With regard to groundwater, the Islamic principles treat it as a communal property with a possibility of private ownership under special circumstances (Chapter 4) (Caponera, 1973; Al-Eryani, 1995). This system addresses the issues of equitable and productive access to groundwater, but fails to provide a regulatory framework to ensure sustainability. Hence, regulating heavy competition for and abstraction of groundwater has become a burning issue.

Before the establishment of the National Water Resources Authority (NWRA) in 1995, no governmental agency was responsible for overall water resources management. Other than NWRA, currently two ministries are involved in water services, namely the Ministry of Agriculture and Irrigation (MAI) and the Ministry of Electricity and Water (MEW). The latter incorporates two general authorities, namely the National Water and Sanitation Authority (NWSA) and the General Authority for

Rural Electricity and Water (GAREW). While the responsibility of MAI is clearly directed to irrigation projects, NWSA is responsible for both domestic water supply and sewerage services in the urban communities, while the GAREW is mainly responsible for water supply in the rural areas.

## 1.2 The Sana'a Basin

### 1.2.1 General overview

The Sana'a basin is located in the central highland of Yemen covering an area of around 3,200 km$^2$ (Figure 1.1). The basin area varies in elevation between 2,200 meter above sea level (masl) in the Sana'a plain to more than 3,000 masl in the surrounding western, eastern and southern mountains. The basin is sloping mildly downward in a south-north direction. The climate of the basin is classified as semi-arid with an annual average rainfall of 230 mm at Sana'a city. The basin area contains parts or all of eleven administrative districts of the Sana'a governorate, in addition to the capital city of Sana'a (the main urban area within the basin). The population of the basin area in 1995 was estimated to be 2.0 million, of which 972,000 (48%) live in the city and the remaining 1.03 million (52%) live in rural areas. It is projected that the population of the basin will increase to 3.4 million and 6.06 million by the year 2010 and 2025, respectively (HWC, 1992c; Foppen, 1996; World Bank, 1997).

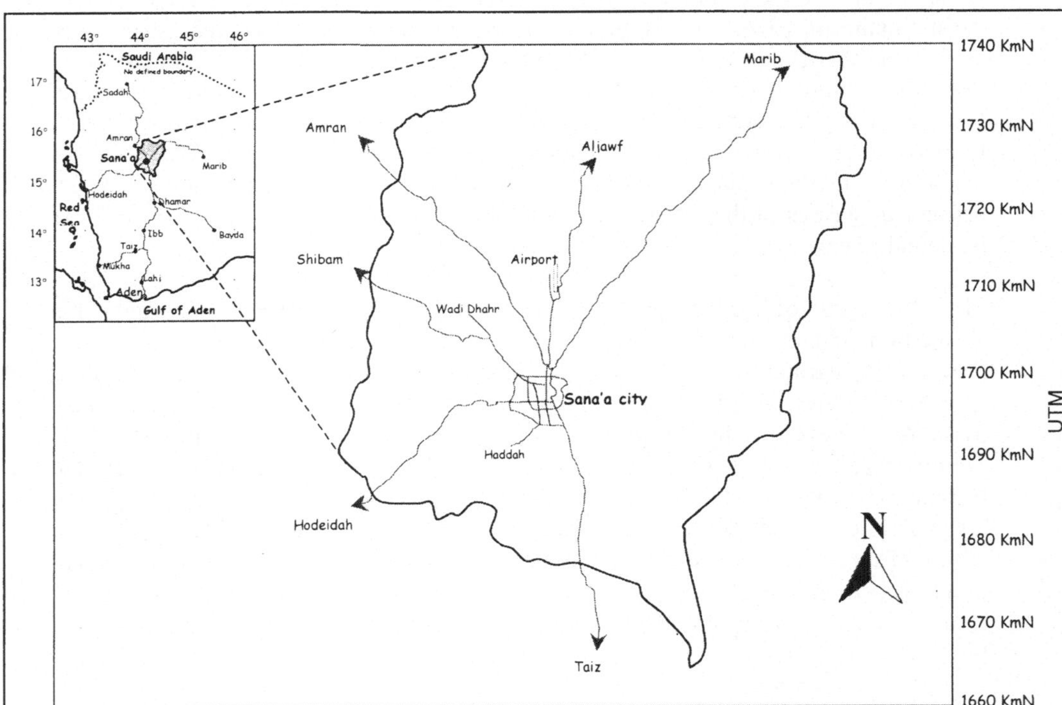


Figure 1.1 Map of the Sana'a basin (after Foppen, 1996).

Land use within the basin falls into four categories: (1) agricultural land (33.2%), (2) urban and built-up land (3.8%), (3) range land (27.2%), and (4) unusable land (rock outcrops) (35.8%). Agricultural land is further subdivided into rain-fed areas covering around 67%, and groundwater irrigated areas covering the remaining 33%. Under groundwater irrigated land, qat[3] is estimated to cover around one third of the total area, followed by grapes and vegetables with 25% each (Bamatraf, 1995).

The heavy reliance on groundwater has resulted in groundwater mining leading to a substantial drop in water levels (2-6 m/a)[4]. Estimates of the natural recharge to the basin vary between 28-60 $Mm^3$/a, with 42 $Mm^3$/a as the latest estimate. Much of the recharge is derived from the intermittent runoff along the wadi courses discharging to the Sana'a plain. Direct recharge from rainfall is limited due to: (1) low rainfall, (2) high soil moisture deficit, and (3) thick unsaturated zone (Italconsult, 1973; Howard Humphreys, 1977; Charalambous, 1982; Mosgiprovodkhoz, 1986; Foppen, 1996). In addition to the natural recharge, there are two other recharge mechanisms currently taking place within the basin area, namely infiltration of domestic sewage (13-15 $Mm^3$/a) (Chapter 2) in Sana'a city and return flows from irrigation (roughly estimated to be 20-25% of total irrigation water application) in the rural areas. The commendable groundwater storage of the basin to a depth of 250 meters below ground level (mbgl) is estimated at 6,050 $Mm^3$, of which around 50% (3,220 $Mm^3$) is considered usable storage (HWC, 1992c).

Although the Sana'a basin is not confined to a single distinctive administrative setup, it is entirely confined to the Sana'a governorate. Furthermore, the basin area comprises entirely or partially eleven administrative districts[5] of the Sana'a governorate (Foppen, 1996). Like in all parts of the country, the tribal structure divides the rural areas of the basin into small tribal regions (villages or settlements) grouped at the highest level into the two major tribes of Yemen (Bakeil and Hashed). It is quite normal to find two neighboring villages belonging to the different tribes. It is usually found that a village (in its entirety) belongs to only one tribe; however, field observations in the village of Sarif shows that due to a long and bloody conflict between its members, this particular village was finally divided into two parts, with each part belonging to a different tribe.

### 1.2.2 Water availability and demand

#### *1.2.2.1 General*

Within the Sana'a basin area all water needs are being satisfied from groundwater resources particularly from the Tawilah aquifer. As of 1995, the extensive groundwater withdrawal has resulted in a deficit of around 120 $Mm^3$/a or 300% of the total natural recharge. The deficit is being satisfied from fossil storage, which is estimated to be in the order of 3,220 $Mm^3$. Assuming a moderate deficit of only 100 $Mm^3$/a and a constant abstraction rate, it is clear that the total usable storage would be depleted within 32 years. Recognizing that groundwater overdraft has been taking

---

[3] Leafs of this tree are chewed as a stimulant.

[4] Reported average decline in groundwater levels are 3.7 m/yr and 4.2 m/yr in the NWSA western and eastern well fields, respectively (Laes and Bamatraf, 1991).

[5] These districts are not water related, but are part of the administrative structure of the Ministry of Local Administration with special connections to the Ministry of Interior.

place since the 1970s, it can be concluded that groundwater in the basin is expected to run dry within the coming few years.

### *1.2.2.2 Urban area*

The city of Sana'a is the only urban center within the basin area, located in the middle of the Sana'a plain (Figure 1.1). The city has expanded rapidly during the last two decades growing from a population of 135,000 in 1975 to 427,000 and 972,000 in 1986 and in 1995, respectively (Central Bureau of Statistics, 1995). Prior to the 1970s, Sana'a obtained most of its water requirements from shallow hand-dug wells. Water was extracted from these shallow wells by pails drawn by humans or animals. In time, an acceptable balance was achieved between withdrawn water and the city's requirements. The water scarcity evolved a dry waste disposal system, with all gray water[6] diverted for garden watering. After the identification of the Tawilah aquifer in 1972 and the establishment of NWSA as an authority responsible for urban water supply and sanitation services, water use in the city increased rapidly. The western and eastern well fields, northwest and northeast of the city were developed in 1976 and 1981, respectively. By 1985, the abstraction from the two well fields was around 9.1 $Mm^3/a$ and increasing to 11.6 and 16.7 $Mm^3/a$ by 1990 and 1994, respectively (Al-Garadi, 1995). Despite the continuous increase in water production by NWSA, the even higher growth of the city has kept the coverage of NWSA supply limited to around 40-50%. The rest of the city's population relies on unregulated private water supplies in the form of small-scale, local piped services and more widely present tanker supply (Al-Hamdi 1994; Keetelaar, 1996; Foppen 1996) (Chapter 2). All private suppliers rely on groundwater from private wells within the city's boundary. Although the background groundwater quality is of acceptable quality that complies with the WHO drinking water guidelines, the risk of groundwater contamination from on-site sewage disposal in the city has lead the public to rely on filtered water from the increasingly popular small unmonitored private filtration units for drinking water (Al-Hamdi 1997) (Chapters 2, 3).

The type of domestic water supply and thus the cost to obtain a desired quantity of "tap water" in Sana'a city seem to be mainly governed by the location of the household. Those who live in areas covered by the NWSA supply and are fortunate to receive high enough pressure, rely entirely on the NWSA supply and thus pay the low subsidized tariff for water. Water, hence, reflects a negligible share in the overall household budget. Those who live within the NWSA supply zone, but at locations where the water service is unreliable (i.e. has inadequate pressure) would have to supplement their water needs through purchases of private supplies, hence raising the share of water in the household budget. The third category comprises those who live outside the NWSA coverage area and are thus forced to rely entirely on private supplies, substantially increasing the water share in the household budget. Depending on the income level of the household, the second and third categories have to pay a comparatively high price for water, which may result in lower demand (Chapter 2).

The leaching of household wastewater into the aquifer, and the contamination of piped water in wastewater logged areas, pose *additional* quality constraints to reliable water supply. People perceive tap water as contaminated and, thus, pay the premium

[6] Discharge water from kitchen uses and washing.

price to purchase qualitatively adequate filtered water in containers for drinking and cooking purposes (Chapters 2, 3).

With no governmental plans to control the expansion of the city, Sana'a is expected to continue to grow at a rate higher than the national annual average of 3.7%. Assuming a moderate growth rate of 5%, the city's population is expected to reach 2.0 and 4.2 million by 2010 and 2025, respectively. Furthermore, assuming a per capita water consumption of 70 l/day, the domestic water need is projected at 45 and 95 $Mm^3/a$ in 2010 and 2025, respectively. Because of the threat of total exhaustion of the available groundwater supplies, the city's population should prepare to face the coming era of expensive imported water.

#### *1.2.2.3 Rural areas*

Although half of the basin's population resides in rural areas, the rural domestic water requirement is not considered critical due to the low per capita consumption (≈ 20 l/cap./d). Agricultural water needs, on the other hand, are comparatively very high exceeding in 1995 the total available renewable groundwater by around threefold. Traditionally, irrigation within the rural areas of the Sana'a basin has been mostly rain-fed with an intricate system of terracing for rainwater harvesting, and using flood diversion systems, thus making the utmost of the available rainwater. Permanent irrigation was only practiced around springs and to a limited scale by water extracted manually from shallow wells (Dubay et al., 1984) (Chapter 5).

Easy access to a reliable irrigation source after the identification of the Tawilah aquifer in the early 1970s and the introduction of drilling and pumping technology promoted rapid expansion in irrigated agriculture, increasing the number of private wells to more than 5,000 by 1995 (NWSA, 1996). Groundwater use for irrigation was developed entirely by private capital, mainly from remittances of migrating laborers in the Gulf States during the 1970s and 1980s. The cropping pattern of irrigated agriculture at the start of the drilling rush was generally a mix of cash crops and cereals. After two decades, the cropping pattern is now dominated by cash crops, particularly qat, grapes, and vegetables (Chapter 5). Cereals on the other hand have been driven out of cultivation due mainly to the government's policy to subsidize imported grains driving the market price of cereals to low levels. In 1990 the area under groundwater irrigation in the Sana'a basin was estimated to be in the order of 18,000 ha, with qat and grapes covering 27% and 16%, respectively, and projected to increase to 48% (10,648 ha) and 23% (5200 ha) by the year 2010. The groundwater irrigation demand within the basin area was estimated at 150 $Mm^3/a$ in 1990 and projected to increase to 290 and 470 $Mm^3/a$ by 2010 and 2025, respectively. The share of qat and grapes in the overall water demand was estimated to be 40% and 24% in 1990, and, if unconstrained projected to increase to 45% and 25% by 2010. The spatial pattern of cash crops cultivation suggests that 80-90% of the total groundwater irrigation water requirements seem to occur within only three districts, namely Hamdan, Bani Hushaysh and Bani Alharith (HWC, 1992c; Bamatraf, 1995).

The rapid expansion in cash crops cultivation was mainly the result of the parallel growth of the city, creating a market with increasing demand, particularly for qat. Moreover, easy access to a reliable irrigation water source (groundwater) coupled with the governmental policy to promote agriculture through major subsidies on energy (diesel) and low custom duties on agricultural related equipment (pumps, generators, pipes, etc.) in addition to the high profits from cash crops, were the major factors that lead to the observed expansion (Chapter 6).

### 1.2.3 Environmental concerns

It is evident that the main threat to the Sana'a basin is groundwater mining, which will probably lead to irreversible damage. This dilemma is not confined to its environmental significance, but will lead to major socio-economic shifts. Future farmers in the basin will look back to the last two decades as their golden age. Dry farming would again be practiced and the low rainfall would probably force a change in the present cropping pattern to be dominated again by crops with lower water requirements, e.g., cereals.

As most of the rural population of the Sana'a basin is involved in agriculture, depletion of the aquifer would mean a collapse of the economic base in the area. With limited employment in agriculture, farmers would probably head for the city in search of employment, thus exerting extra pressure on the city's services, particularly the already fragile municipal water supply.

A second major environmental issue is the on-going contamination of groundwater from urban sewage disposal[7]. Although such contamination poses a health hazard when contaminated groundwater is pumped up for domestic uses, at the same time the wastewater disposal has contributed considerably to groundwater recharge. In other words, if domestic sewage is not allowed to infiltrate to the underground, while private wells continue to supply half of the city with water supplies, groundwater levels would certainly drop faster and probably reach a point of total depletion. Thus, there would be no groundwater to contaminate while all the sewage would be "wasted", i.e. evaporate and used for irrigation of low value crops (Chapter 7).

## 1.3 Groundwater as a Common Pool Resource (CPR)

### 1.3.1 Definition, types and problems of Common Pool Resources

Ostrom (1990) defined CPRs as "a natural or man-made resource system that is sufficiently large as to make it costly (but not impossible) to exclude potential beneficiaries from obtaining benefits from its use".

There are similarities as well as differences between CPRs and public and private goods. A CPR and a public good, such as sunshine, resemble each other in the difficulty to exclude beneficiaries from their use. The main difference, however, is the CPR's subtractability: the resource system of a CPR is affected by the withdrawal of resource units, while that of a public good is not. The similarity between a CPR and a private good is the subtractability of the resource units, while their difference lies in the excludability of a private good: unlike CPRs, it is relatively easy and inexpensive to exclude others from the use of private goods. Groundwater and irrigation water are typical CPRs. Water from a piped water distribution is a private good, although illegal connections, frequent in cities in many developing countries, suggest that it can assume to some extent CPR characteristics.

---

[7] Groundwater contamination has been detected in the city resulting from on-site disposal of sewage, as well as along the effluent channel of the ponds, north of the city (Al-Hamdi, 1994; Wiesenekker, 1996)

Ostrom differentiated between two aspects that characterize a CPR, namely the resource system (stock) and the resources units (quantitative flow). While the resource units of a CPR are extractable and thus subject to private ownership, the resource system is subject to joint access by all users. For instance, in the case of groundwater, the aquifer is the system and the abstracted water, a discrete, more manageable quantity, are the "resource units". Both aspects may be governed by different ownership, allocation and management rules, as they may be perceived by the users as different entities. In addition, Alaerts (1996) argues that resource management issues often need to be distinguished from infrastructure management issues. In irrigation systems, for example, the joint financing, operation and maintenance by the users of the infrastructure calls for different institutions and capacities than the resolution of ownership and allocation of the water resource. CPRs are also distinguished by their access: open access and limited access CPRs. While open access CPRs have no limits on who can use the resource, users of a limited access CPR are usually a well defined group. Open access is considered the "extreme case" where entry and production are open to all parties and usually associated with undefined property rights and pronounced competition. In limited access CPRs, property rights and the existence of allocation, entry and production rules, and thus conditions that moderate competition, may be present.

Resource economics showed that the total amount of resource units withdrawn from an open access CPR by independent users is greater than the optimum economic level (Clark 1976; Dasgupta and Heal, 1979). The competition for a CPR will lead users to attribute less value to benefits that they expect to receive in the future and more value to present benefits. With undefined property rights and non-exclusive access, the "rule of capture" usually governs the use of the resource stocks and may lead to complete destruction of the resource. Over-exploitation of a resource will usually result in rent dissipation[8]. The absence of secure property rights over the resource units will limit market exchanges result in lower value uses of the resource. By equating only their private gains and costs (neglecting collective and long-term benefits), users have an incentive to exploit the resource too rapidly. Competitive use over resource units of CPRs results in externalities when the use by some users reduces the use opportunities of others. This is, for example, a common feature of groundwater over-pumping, as heavy abstraction by one pumper usually results in a decline of the water table in his neighbor's well, forcing him to incur higher costs. Shah (1993) suggests that, although pumpers usually have enough information to predict such externalities, they fail to internalize these externalities due to the absence of social, behavioral or other regulatory mechanisms to restrain collective pumping thus leading to socio-ecological crises.

Given the short-term value of water in a water-short environment, one can expect that pumpers would respond to the absence of enforceable rules on access and use by independently taking the rational action to over-pump, thus creating higher demand on the resource. Empirical cases of over-exploitation of groundwater are found all over the world (India, Yemen, Jordan, USA, etc.) thus supporting such expectations (Beaumont, 1985; Moench, 1993; Shah, 1993; Al-Sakkaf et al., 1999; Macoun and El-Naser, 1999). Without a fair, orderly and efficient method to allocate and distribute

[8] Rents are dissipated when the marginal returns from use are smaller than the marginal cost of appropriation, and occur when too many users are allowed to withdraw more than the optimum amount of resource units.

the resource units, local appropriators have little incentive to contribute to sustainable use of the resource system.

### 1.3.2 Methods to govern CPRs, regulation vs. privatization

There are mainly two methods to manage CPRs, namely external (state control) and internal (by self-organization). In a situation where officials indicate that it is their responsibility to manage and solve CPR problems, beneficiaries of the CPR would probably not initiate an action, but rather wait for the central intervention to solve the problem. In other situations where individuals are encouraged to participate in the management of their affairs, self-organization of CPRs becomes a likely scenario. For individuals to engage in institutional change, the prerequisite is that they feel a need for such change. In other words, there should exist a driving force such as a clearly visible, high level of externality resulting from their appropriation of the resource units that will eventually threaten their welfare, leading them to believe that collective action will produce higher collective and personal benefits. Analysis of successful self-organization on groundwater basins in California, USA, shows that the institutional change was only initiated when severe and obvious externalities (decline in water levels and sea water intrusion) were threatening the future-benefit-generation capacity of those basins (Blomquist, 1992).

Some scholars have suggested that the state should control most natural resources to prevent their destruction. Others have argued that privatization of natural resources is able to solve the over-exploitation problem. Observations, however, suggest that neither the state nor the market is consistently successful in sustaining long-term productive use of natural resources. Governmental regulations of groundwater development, for example, have proven ineffective in some regions of India. Public policy instruments were perceived in India as inefficient and un-equitable, and especially disadvantageous for the poor (Shah, 1993; Moench, 1994; Shah, 1997). Saleth (1996) viewed indirect governmental measures to manage groundwater through a power tariff to be an ineffective instrument to simultaneously achieve the three goals of *equity* (in access), *efficiency* in terms of agricultural productivity and *sustainability*. However, in the successful cases of self-organization of CPRs (Ostrom, 1990) appropriate rules take into account the specific attributes of the physical system as well as the cultural, economic and political setting of the local conditions. While locally devised rules are usually tailored to the specific situation and are normally based on the concurrence of the local stakeholders, rules imposed by external sources may be less effective, or even counterproductive.

State control of resources does commonly not manage to take into account the variations in local conditions, which leads in many instances to unsuitable rules. However, complex allocation problems arise with privatization of the resource system. Defining transferable property rights to the flow of the resource units (pumped water) does not mean privatization of the resource system itself (the aquifer). Shah (1997) demonstrated that in water-abundant areas, water markets (formal or informal) are socially beneficial, especially when such markets are competitive and efficient, but that by contrast, in water-scarce regions suffering groundwater over-exploitation, competitive water markets contribute to resource depletion. Shah suggested that the environmental dis-benefits of water markets in such fragile situations will rise proportionally with their efficiency and equity benefits.

Shah (1997) argues that the emergence of informal groundwater markets in India, Pakistan and Bangladesh have resulted in positive and significant efficiencies and equity, in the form of more reliable income for non-borehole-owner farmers, higher utilization efficiency of abstraction equipment, and wage improvement for the landless. Saleth (1996) on the other hand, considers these informal institutions to be inefficient and inequitable and views their positive efficiency and equity gains to be insignificant as compared to their fundamental inequity, both intra- and inter-generational, in addition to their unsustainability.

Although informal water markets have evolved in many countries, in the absence of clear water rights it has been argued that the water sellers (or the borehole owner) sell only the *service* of their extraction equipment. Therefore, one could think of the inability of buyers to invest in their own abstraction equipment (borehole and pumping equipment) as the driving force behind these "surrogate" markets, which would thus disappear if most water buyers overcome this financial constraint (Shah, 1997). Although the situation with regard to water rights in Sana'a resembles that of South Asia, comparatively lower market activity in Sana'a is the result of the structure of borehole ownership with its high level of ownership sharing (Chapter 5). Although selling and buying of water is taking place in these "surrogate" markets, water is not treated as strictly an economic good. For water to become a "true" economic good, the underlying structure of property rights to water would have to become more clearly defined and enforceable, with tradable rights[9].

In order for markets to emerge and become effective in raising the value of the resource units, property rights need to be clearly defined. This usually takes place once boundary and authority rules are identified and enforced. As boundary rules identify the conditions to access the resource, authority rules specify the allocation structure of the resource units to the different appropriators. For sustainable use of a resource, authority rules need to be devised in accordance with the safe yield of the resource. Satisfying the additional criterion of tradability of the rights increases the possibility for a shift to higher value uses of the resource system.

### 1.3.3 Models of CPRs and collective actions

Three theoretical models try to explain the individual's behavior related to the use of a CPR: (1) Hardin's tragedy of the commons, (2) game theory such as described in the Prisoner's Dilemma, and (3) Olson's logic of collective action. These models lead to the prediction that depending on the individuals' expected private benefits, users of a CPR will or will not cooperate to achieve collective benefits. Without information about the possibility of collective gains, individuals are perceived to be trapped and unable to change the rules affecting their incentives. The incentives and choices of users of a CPR where there are no restrictions on access or use, are very similar to those faced by players in a prisoner's dilemma game. Each user's dominant strategy is to exploit the resource without willingness to contribute to its preservation or maintenance. Ostrom limits the prediction capacity of these models to situations of large-scale CPRs, where the communication level between users is low, every user acts independently, no attention is given to one's actions, and the cost to change the situation is very high. Ostrom's conclusions are derived from the fact that these

[9] For the case of groundwater, tradable water rights would imply at least the separation between water and land rights. For further discussion on allocation mechanisms refer to Dinar et al. (1997).

models fail to explain the empirical cases, where through self-organization users of CPRs have been able to change the institutional setup and achieve a collective action so as to generate additional benefits and/or sustain the use of the resource. It has been also argued that although these models suggest a low cooperative capacity of individuals to reach consensus to achieve a collective benefit or prevent a collective harm, they do not specify the likely course of action of the individuals when given the chance to craft their institutions, nor do they clarify the capacity of individuals to develop institutions that can lead to better outcomes.

In trying to empirically explain successful cases of self-organized CPRs, Ostrom was limited to relatively small-scale[10] CPRs under conditions of substantial scarcity. Ostrom's objectives were to identify (1) a set of design principles for successful self-organized cases, (2) the incentives that incite individuals to sustain self-organization, and (3) the external and internal factors that enhance or impede the capability of users to govern CPRs. Instead of assuming that all CPR situations involve only one underlying structure (such as that of the prisoner's dilemma), it was assumed that users face a variety of local appropriation[11] and provision[12] problems to be solved. The key issue is how to allocate a fixed, time-independent quantity of resource units so as to avoid rent dissipation, while at the same time reduce the conflict and uncertainty over the distribution of such rights.

### 1.3.4 Tasks and design of institutions

To manage CPRs whether through state control or self-organization, common challenges emerge: (1) coping with free riders[13], (2) solving the commitment issue, and (3) monitoring and sanctions to enforce compliance to shared rules. Generally, the prime task of institutions lies in adjusting the incentive system of the appropriators, so that free riding is no longer the preferred dominant strategy. To clarify this, it is useful to examine the factors that contributed to the success in some studied cases. Ostrom (1990) identified two categories of factors, namely internal and external. Internal factors relate to (1) the capacity of users to communicate, (2) the level of trust between the users, (3) the shared sense that users face a common future, and (4) the absence of strong opposers (those who stand to lose private benefits as a result of the collective action). If a situation is characterized by a low commitment of individuals, low level of trust between the users, no strong sense that a common future is shared, and the presence of influential parties opposing institutional change, collective action is unlikely and would have to be facilitated by external assistance. External factors include (1) the level of autonomy of the users, and (2) the existence of incentive systems resulting from official policy. CPR users with no or low autonomy to initiate institutional change, and who are being encouraged to over-exploit the resource system through official incentives (e.g., subsidies) are unlikely to succeed in establishing an institutional change able to sustain the use of the resource.

Changing a CPR situation from one in which users act independently to one where they adopt coordinated strategy to obtain joint benefit or reduce joint harm does not

[10] CPRs that are entirely located in one country with 5,000-15,000 users who depend on the CPR for their welfare.

[11] Related to the allocation of the resource units.

[12] Related to the maintenance of the resource system (stock).

[13] Free riding is taking advantage of the compliance of others (e.g., a farmer who uses an irrigation canal without contributing to its maintenance).

occur automatically and is not likely to be a smooth process, especially when important heterogeneity exists among the negotiating parties, or within each of these parties. Moreover, such process can be very costly if it requires different types of accurate information. Sequential bargaining is a characteristic of collective action in CPR problems, with each round involving considerations of new information on aggregate and individual costs and benefits. It is important for successful bargaining that the influential parties see their private welfare improved by the collective action in order for them to participate, otherwise they will decline to participate even when important social or collective gains are expected from the collective action.

Ostrom (1990) concluded that seven design principles are essential to craft robust self-organized institutions: (1) clearly defined boundary, (2) symmetry between boundary rules, authority rules and the local conditions, (3) collective choice arrangement with participatory approach in rules modifications, (4) effective monitoring, (5) graduated sanctions, (6) mechanisms for conflict resolution, and (7) supporting governmental and political regime.

Libecap (1995) indicated two stages in the bargaining process to reach a collective action and solve CPR problems: (1) the discussion of the aggregate gains of the proposed institutional change, and (2) the resolution of distributional problems of these gains. Libecap suggests that the first stage is usually clear to the involved parties, particularly in the case of non-renewable CPRs. The second stage, however, consumes more time and it is only after the quality conditions of the resource have deteriorated badly that an agreement becomes possible. An exception to this generalization is in newly discovered CPRs such as a newly discovered aquifer, where an advantage lies in the fact that no prior production has taken place[14], resulting in higher homogeneity of the involved parties, and leading to a relatively quick progress in reaching collective action.

### 1.3.5 Role of external (official) involvement

In the cases of groundwater basins in California, USA, cited by Ostrom (1994), the role of external assistance was a key factor in establishing institutions capable of dealing with CPR problems, and in the majority of these cases actually succeeding in overcoming the problems. The government's involvement was commonly not in the form of devising the rules themselves, but in the provision of facilities and mediation that enhanced the ability of the users to engage in institutional change. These facilities were in the form of (1) allowing a high level of autonomy, (2) provision of good technical information regarding the resources, (3) providing negotiating and conflict resolution arenas, in addition to (4) bearing some of the costs of the process. Ostrom doubts whether the pumpers in the groundwater basins in California would have been successful had it not been for the information services provided by the state and federal departments and agencies. The role of the political regime in California was not only confined to the provision of facilities, but included the oversight by the local and state officials to supervise and ensure the equitable solutions. Although the governmental intervention in the California groundwater cases might be viewed by some as prescriptive, it cannot be considered as direct central control. Nevertheless, the failure of groundwater pumpers in the Mojave region to achieve a similar positive

[14] The absence of past gains lowers the conflict potential of distribution.

outcome to that in the neighboring basins illustrates that even when the right supporting political environment exists, successful resolution is not guaranteed.

### 1.3.6 Rules, types and functions

In all cases of successful management of CPRs, rules that severely constrained the actions available to the users have been established. Ostrom identified two general types of rules supported by two additional criteria that exist in all successful self-organized CPR cases. The two rules are boundary and authority rules, while the two criteria are the presence of an active form of monitoring and sanctions, and the absence of a "grim trigger" strategy[15]. Devising boundary rules is a difficult task and usually faces resistance from the users of the resource. Boundary rules clarify the criteria under which an individual is able to access the resource, which may be based on land ownership, historical use, or shared ownership. Without the complementary role of authority rules, boundary rules alone are not sufficient to overcome CPR problems. Nevertheless, they are an essential first step in the process. Authority rules are those connected to the allocation of the resource units, which may be limited by space, time, or quantity. In the California groundwater basins, the basis of the authority rules was the sustainable yield of the basins, which was translated into proportional reductions in pumping for all pumpers according to historical withdrawals. Having specified the allocation of the resource as a certain allowable withdrawal rate, water rights were clearly defined and separated from the land, thus leading to market exchanges and a shift to higher value uses of the groundwater. Moreover, volumetric abstraction charges were established to generate funds for maintenance and administrative purposes.

Having devised boundary and authority rules, commitment by the users to follow those rules is not automatic. If everyone abides by the devised rules, the flow of resource units can be more predictable and efficient, leading to a sustainable resource system and lowering the level of potential conflicts. Nevertheless, in the presence of substantial private benefits, the temptation to break the rules is high, especially given the uncertainty of compliance by others. In other words, no one wants to follow rules that everyone else is breaking. To solve the commitment problem, users need to motivate themselves to monitor and be willing to sanction violators. When users adopt a contingent strategy for compliance with rules, they become motivated to monitor each other. It can thus be generalized that without effective monitoring and sanctioning, commitment cannot be achieved, and without commitment the efforts to device new rules are in vain. In the groundwater basins in California, monitoring is conducted by an independent neutral water master, who collects specific information and produces an annual report publishing the individual pumping rates[16], cases of non-compliance, and the state of the resource. This report is made public and is widely available to all pumpers, ensuring a low level of disparity of information, which is viewed to be a key factor in reaching credible commitment to follow the rules. The entire process, even with reliable public information and effective monitoring, can collapse in the absence of effective sanctioning to deter rule breaking. Sanctions should not only be declared and accepted by the users, but they should also be fair, practical and enforceable. Graduated sanctions provide flexibility, where

[15] Grim triggered strategy is one where a rule abider decides to break the rules upon knowing of rule breaking instances.

[16] Obtained from direct reading of installed water meters.

minor violations (e.g., at times of emergencies) would only involve small punishment, increasing to different levels according to the type of the violation. It is also important that information on rule breaking and applied sanctions are reported to the involved individuals so as to inform and assure all users of the level of compliance (Runge, 1984; Wade, 1988). Monitoring and sanctions are necessary to keep the rate of compliance high enough to avoid triggering a process of increasing rule infractions.

It can thus be concluded that in order to overcome CPR problems, reliable information, not only on the physical attributes of the resource (boundary, yield, etc.) but also information related to the use of the resource and the level of compliance, are to be supplied and maintained. The cost to obtain such information depends on the situation, and is based on the accuracy and frequency of the needed information.

### 1.3.7 Influence of size and number of users

The previous discussion has concentrated mainly on the understanding of the incentives leading to collective action to solve small-scale CPR problems. A larger scale of the CPR increases the complexity of the problems by introducing larger uncertainties and greater heterogeneity among the users. With higher numbers of individuals involved, the sense that problems exist sharpens and the potential for conflict increases. Moreover, large CPRs involve areas with different levels of scarcity, thus lowering the commitment level of users in areas with a lower level of scarcity[17]. Failure to reach an agreement in the Mojave region, California, may be explained by the complexity resulting from the larger size area compared to the more successful smaller basins. Although the initial steps taken to initiate a change in the institutional arrangement in the Mojave region were similar to those in the other basins, major difficulties were encountered in the establishment of boundary as well as authority rules due to conflicting views on (1) definition of the region (one vs. several basins), (2) scale of the problem (local vs. regional), (3) basis for distribution of the resource units (historical vs. coequal rights), (4) parties to be involved (large users only vs. all users), and (5) allocation of the resource (rights associated with vs. separated from land). The divergent views on these issues, which are mostly attributed to the large size and greater number of involved parties in the Mojave region, lead to conflicts and finally suspension of the litigation.

### 1.3.8 Conditions for institutional change

There are six conditions that contribute to a higher chance of institutional change to solve CPR problems: (1) the homogeneity of interests of the users, (2) the availability and accuracy of relevant information, (3) a small number of involved parties (individuals), (4) the discount rate of the use of the resource units, (5) the willingness of the users to adopt voluntary reciprocity, and (6) the associated transaction costs. The question that rises then is: when will individuals think it is the right time to peruse a solution to resource over-exploitation? Ostrom suggested in optimism that collective action could be taken at an early enough stage to save the resource system, and used the success of empirical cases to support such position (Ostrom, 1990).

[17] In the case of the west groundwater basin in California, the city of Hawthorne was not in favor of an institutional change and adopted a held-out strategy for many years due mainly to (1) the city's location, as city officials believed that their water supply was relatively safe from sea water intrusion, and (2) the city persistently viewed its need to serve the municipality with water supply as superior to the needs of other users in the area.

Libecap, on the other hand, was more pessimistic suggesting that for many local CPRs effective collective action will not occur until major losses to the resource have already been observed. He argued that serious dissipation of resource rents is required first, with the value of the user's share under the status quo declining more and more until all users see themselves as gaining by the collective action. During this process (over-exploitation stage) the number of individuals declines as some users are unable to stand the increasing externality costs and are pressured to exit production. With the remaining individuals facing mounting externalities, they become more homogeneous in their future production capabilities (Libecap, 1995).

CPR problem resolution typically follows a sequence of stages, in which the competing resource users evolve from competition to collective action (Alaerts, 2000). During this process, the role of *knowledge* and *information* features prominent:

Stage 1. Resource is plentiful compared to use rate. Unrestrained competition among resource users.

Stage 2. Resource shows signs of stress or depletion. Competitive users suffer from unrestrained competition (increasing cost to acquire same resource quantity, or less quantity available). By heuristic approach (trial-and-error), users increasingly *understand*

1. the physical attributes of the resource system (groundwater, fisheries, forest, etc.) which suggests them ways to maximize income under competitive conditions.
2. the behavior of their peers and the nature of the competition.

Stage 3. Users make rational assessment, *understand* that they stand to win from cooperation, and take decision for collective action.

Stage 4. "Free riders" reap benefit by not abiding by the joint agreement and abusing restrained resource use by the others. The users collective, therefore, imposes *transparency* on decisions and actions of the individual resource users, and *enforces* the agreement by penalizing deviations.

Significantly, in stages 2 to 4 an outside and "objective" facilitator can play an important role, by (1) helping to acquire the fundamental information and knowledge, which is of especial relevance in the case of physically complex CPRs and in cases where "win-win" results may be difficult to discern, (2) brokering negotiations, and (3) take part in the monitoring of the agreement and act as holder and disseminator of information on the behavior of the respective parties to the agreement.

### 1.3.9 Possibility of collective action in Sana'a

The previous discussion allows to better assess the feasibility of a collective action to overcome groundwater mining in the Sana'a basin. The only boundary rule that seems to exist in the Sana'a basin is the limitation of the pumping rate to owners and shareholders of existing wells imposed by the well's and pump's capacity. Nevertheless, the lack of a rule that restricts the number of wells renders the existing rate control rule ineffective with respect to the criterion of the sustainability of the resource. Although access to groundwater thus is restricted to some extent to the local rural landowners and farmers, the large population dependent on groundwater in the basin has resulted in a rapid depletion. Even though, in theory, the Sana'a basin case may be characterized as a limited-access CPR, the large deficit between demand and renewable supply shifts the scale of the problem to that of a more complex open access CPR.

It is unlikely that groundwater users in the Sana'a basin area will initiate actions to establish self-organized institutions to deal with the groundwater over-exploitation because of the following factors:

1. The government has assumed a degree of responsibility by announcing that management of water resources falls under the authority of the relevant official agencies[18].
2. The large size of the basin area increases the complexity of the problem and introduces higher uncertainties and greater heterogeneity among the pumpers. Furthermore, the existing complex tribal structure within the basin area significantly lowers further the level of trust between user groups.
3. The available information regarding the state of the resource is not accurate. Importantly, the available information has not been widely distributed among the users. In addition, the perception and views of groundwater users towards declining water levels are not based on scientific of physical insights (Chapter 4). Acquiring and distributing reliable information is very costly (given the complex geological setting) that neither the government nor the local users are willing or considering to finance.
4. Most users face high discount rates with groundwater being critical in their main economic activities and welfare.
5. The existing political regime is perceived by the water users to be unable to provide the facilities and support to the local pumpers in order to initiate an institutional change.

An equally important factor is the timing for an institutional change as related to the degree of progress in the resource exhaustion. Unfortunately, the large capacity of the Tawilah aquifer allows gradual depletion imposing only slowly increasing externality costs to the pumpers, that can be offloaded to the local market prices. However, this slow rate of change averts a perception of urgency and preparedness in a social environment of low literacy, poor information and low cultural homogeneity. Once near depletion is reached, the system shifts to very rapid "catastrophic" change and excessive externality costs to which the local economy cannot any longer adjust. This scenario corresponds best to Libecap's pessimistic model (see above). Although farmers at present are experiencing externalities in the form of inter-related decline of water levels resulting from over-pumping, these externalities are not yet viewed as a major cost. The additional cost incurred by the farmers from declining water levels arguably has been low, compared to the generated profits (Chapter 6). It can be assumed that once the externalities become severe so as to create a strong enough incentive for pumpers to seek a solution, either external (by asking the government for assistance) or internal (by initiating a self-organized institutional change) collective action to limit pumping becomes possible. Nevertheless, given the profitability of irrigated agriculture, it is doubtful that farmers would reach this stage, since total depletion of the aquifer may take place before collective action to control pumping is reached[19].

---

[18]The constitution declares all natural resources to be state property and establishes them to fall under state authority.

[19] In some parts of the country, qat cultivation is profitable even when irrigated with tankered water.

## 1.4 Cost, price and value of water

CPR theory helps to describe the competitive as well as the subsequent collaborative behavior of water users. This behavior is based on awareness (insight, information, knowledge) that is shared and causes the actors to seek win-win situations in a longer term perspective. As such, the theory embodies the concern for the resource sustainability. It allows to analyze and quantify the incentives for decision-making that pertain to longer term resource management. The time scale involved is typically that over which the resource system could get depleted or damaged to the point that the users stand to lose.

In the short term, water user behavior is subject to different incentive systems with a shorter time horizon. It is increasingly being argued that individuals behave semi-rationally in an economic sense (for a good review, see e.g., *The Economist*, December 18, 1999, p. 69-71). In other words, the main incentives are of a financial (or economic) nature, but individuals often *perceive* their benefits, losses and risks in more complex ways, and some parts of their utility may not be converted in financial terms. In the context of competition for exhaustible groundwater, for example, people tend to underrate risks of rare occurrences (such as floods or natural disasters, or aquifer exhaustion) and overrate those of regularly occurring ones. Similarly, people are "loss averse" in the sense that they would value lower the $100 they would gain, than that they would attribute value to $100 they would lose (Kahneman and Tversky, 2000). Arguably, farmers' behavior on the Sana'a plain is influenced by such perceptions.

It is assumed here that the financial net benefit or gain to the water user is one key incentive, probably the most significant one to determine groundwater use in an agricultural setting, and choice of water for domestic uses in an urban setting. The net benefit represents the *value* of the water to the user. In a broader sense, the behavior of institutional water users, and of the national economy with respect to water use, are similarly subject to the water's perceived value (Alaerts, 1996)(Figure 1.2). To the nation this value can be expressed in monetary terms per unit of water, or in other units such as, for example, jobs supported per unit of water. In economic terms, proper valuation should value polluted water negatively as it decreases the utility of the water resource. In a context of sustainable water use, the value of water should be set at at least the cost to restore the resource system to the situation before its use. In the case of groundwater, for example, this would imply that the value of a unit of groundwater should equal or exceed the cost to replace the abstracted amount with an equal volume of water of similar quality.

The *price* of the resource units functions as one of the main allocative mechanisms. High price fosters saving water, whilst low price tends to remove quantitative in-built restrictions on water use. In a seller's market (an arena with many purchasers but few suppliers) the price tends to be set by the maximum the seller can obtain, i.e. by, on one hand, the desire to maximize profit, ensure enterprise financial sustainability, etc., and on the other hand, by the willingness of the buyer to pay for this commodity. In a buyer's market (an arena with few buyers compared to the number of sellers) competition tends to lower the price to a level close to that of the real cost. Other mechanisms that assist in resource allocation are the trading of water use rights, prescriptive regulation, and various forms of consensus and covenant systems. Among the latter, local traditional, tribal and religious rules often define the boundary and authority rules for the CPR. In the case of groundwater, the abstraction rights are often defined by the well ownership rules (Chapter 4).

The *cost* of the resource unit consists at least of the direct costs to produce the commodity, namely the capital and the recurrent cost for operation and maintenance. Depending on the scope of the analysis, other costs can be included, such as those for acquiring capital and insure against risks, and opportunity costs. In case of institutional behavior, the cost can also include transaction costs.

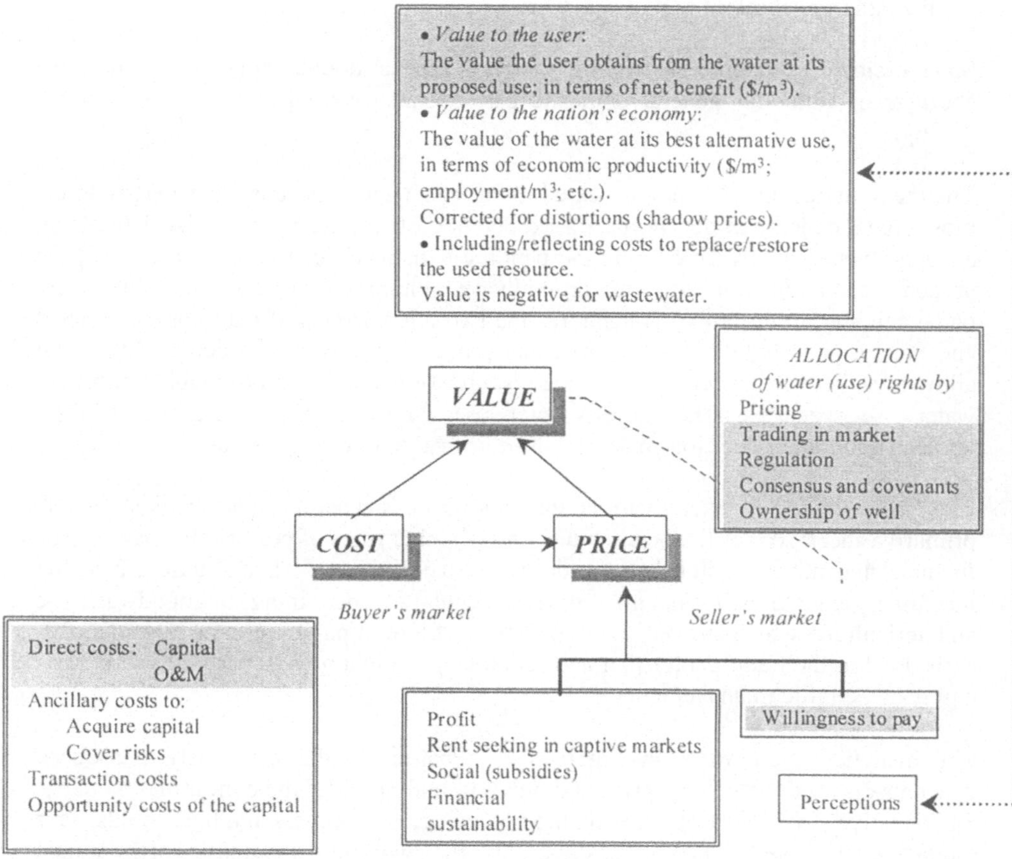


Figure 1.2 Value, cost and price of water define a water user's behavior (Alaerts, 1996). Shaded areas pertain to the factors considered in this study.

The cost is an important determinant of both the price and the value of the commodity, but in an unregulated seller's market the price may be more driven by external factors and in effect bear little resemblance to the true cost.

Groundwater pumpers in Sana'a's plain as well as city inhabitants are to decide daily on their water use pattern. To a large extent these decisions are likely to be decided upon by the financial incentives.

## 1.5 Analytical framework

The Sana'a basin is a unique setting, or arena, in which the rapid depletion of the groundwater stock is caused by four sets of simultaneous competitive arrangements:

1. Farmers compete among each other;
2. The aggregate of farmers competes with the citizens of Sana'a;
3. Citizens compete against themselves to use the same groundwater aquifer for both wastewater disposal (leaching cess pits) and for drinking water abstraction;
4. The current generation competes fiercely with the future generations (and because the latter are absent, the first will win).

Sana'a citizens, in contrast to the farmers, face a double concern: quantitative shortage of water for general domestic use, but also shortage of water of potable quality.

The main interest in this study is geared to the sets 2 and 3, because they pertain to the most urgent concern of the short-term exuberant costs for the Sana'a citizens to obtain adequate water supply, due to (1) the pending exhaustion of the aquifer, and (2) the gradual contamination of the groundwater pumped for domestic uses. By understanding the incentive systems for the two main groups of water users (farmers and Sana'a households) it is assumed that process options can be depicted that will allow to achieve better regulation of the resource or the definition of viable alternative water use scenarios. The largest water user group is the farmers, and, thus, understanding their decision-making will be critical to devising a solution.

The basic approach taken here is that all users' behavior is determined by (1) primarily, incentives of financial nature, and (2) other rules of prescriptive nature. The financial incentive structure can be described using the model of Figure 1.2. Other incentives relate to the formal governmental regulations pertaining to groundwater use and agriculture, which are very weak in Yemen. More important are the informal ones embedded in the traditional, tribal and religious rules that govern access to resources, equity and conflict management.

The analytical framework in Figure 1.3 is based on the previously introduced concepts on water value/cost/price and on CPR, and provides the conceptual model of the incentive system that operates on the two key actors: farmers and households. This model at the same time is designed as the working hypothesis. The model distinguishes the retained independent variables. The study limits itself to the items shaded grey in Figure 1.2. For the farmer, the independent variables are the cost to abstract water, the market value of the produce, the traditional and informal rules on water use and well ownership, the formal governmental rules on water use, and the long-term value of water associated with the awareness of unsustainable mining of the CPR. The households are subject to a similar set of variables: the water quality concern that is yet to be quantified, and the long-term value of water associated with the unsustainable mining of the CPR. In addition, the overdraft by the farmers causes concern about domestic/drinking water quantity.

On the farmer's side, the incentives influence decisions regarding strategy on crop choice, (hence) strategy on water use, and (un)willingness to resolve the CPR competition. On the household's side, the incentives influence decisions regarding the "water basket" to be purchased to mitigate risks, the saving of water, the prices to be paid to obtain the desired quantities of water of specified quality, and the (un)willingness to resolve CPR competition. A question of particular interest is that of the feasibility of re-use of municipal wastewater in irrigated agriculture in order to trade high-value groundwater of drinkable quality from the farmers for lower-value water of a quality that is not potable but still suitable for agriculture.

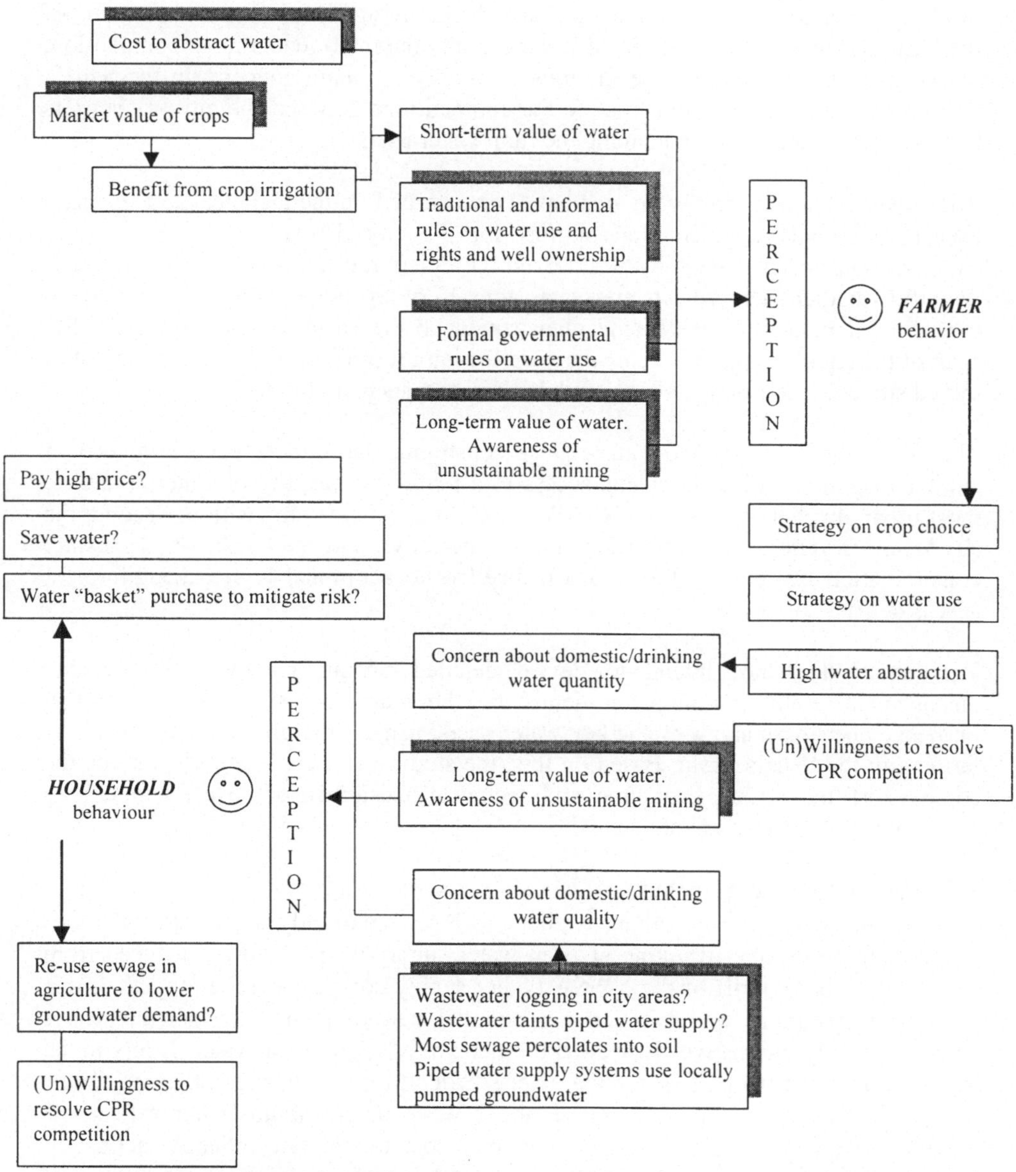


Figure 1.3 Determinants in the incentive systems that influence water use behavior of farmers and Sana'a citizens in their competition for groundwater. Shaded: independent or basic variables.

This study will be organized along the lines of this analytical framework, and address in the subsequent chapters the issues mentioned in the boxes. However, the limitations of the study did not allow to comprehensively analyze all factors.

## 1.6 Research questions, approach and structure of the study

### 1.6.1 Research objectives and hypotheses

The population in the Sana'a basin area (both urban and rural) are facing a severe water crisis in the near future. Groundwater, the exclusive water resource, is being mined at a rate that threatens the availability of city tap water, the availability of drinking water to the urban population and the entire agricultural economic basis of the basin. Generally, in situations of water scarcity the inter-generational competition, or access to water by future generations is usually the main concern. In the Sana'a basin, however, the first concerns are the competitions between the city water users and the rural water users, and among the rural water users.

Prior investigations in the basin area have concentrated on the technical and financial aspects of supply augmentation options for the city. However, the options are expensive and need large subsidies. Given the deteriorating economic situation it is doubtful whether such projects are sustainable. Very few attempts have been made to assess the applicability of demand management in the basin area to extend the life span of the aquifers and thus delay the need for import projects, or to devise long-term mixed strategies combining supply and demand management tools.

The urban dweller faces two water-related constraints that impose upon him costs of similar magnitude. First, he wants access to a minimum quantity of water, which to most urban dwellers (not served by NWSA) is an expensive proposition. Second, the *de facto* recycling of wastewater under the city, and the risk of tap water contamination by sewage, forces urban dwellers to additional high expenditure for access to drinkable water.

This study does not intend to devise detailed operational tools for demand management. Rather, it aims at identifying the main incentive frameworks that influence decision making by the key water users, namely the urban dwellers and the farmers in the Sana'a basin. Based on this, the study will then attempt to outline the contours of the social, economic and technical feasibility of several options to augment supply and/or adjust demand.

The research hypotheses include:

- In aggregate, the urban dwellers of Sana'a currently spend several times more on their drinking and tap water supply than would be deduced from the basic tariff of the NWSA (water supply company). Taken together, this expenditure would cover a substantial part of a more sustainable water supply alternative. However, the net overall charge for water differs much among city quarters making it very difficult for city dwellers to arrive at one common strategy. Therefore, the NWSA tariff should better reflect the overall water's opportunity and market cost, to facilitate collective action.
- Agricultural water uses behave rationally in the sense that they adjust their water use to financial incentives (earning on the market of qat and grapes). On the other hand, the farmers behave irrationally, in the sense that they do not use water solely to maximize their earnings, and that they semi-consciously evade the issue of the sustainable use of water.
- The existing traditional and *shari'ah* inspired ownership rules and water use rights are effective to realize the economically productive value of the

groundwater and to achieve equitable distribution among tribe members, but they fail to outline a strategy for sustainable water use.

- Wastewater from the city can be re-used in agriculture at comparatively low cost, and thus be exchanged with farmers for access to groundwater of good quality.

### 1.6.2 Approach

To attain these study objectives secondary data were collected from (1) reports of prior groundwater related projects especially with respect to the basin's geo-hydrology and the NWSA's water supply system, and (2) extensive review of relevant literature. The most important source of primary social economic data has been collected from the "field", i.e. farmers and the city: (1) a representative farm was monitored for three consecutive years, and (2) two broader field surveys were carried out, of which one was conducted in the rural areas covering the regions with the highest groundwater abstraction rates in the basin, and the other in the city to evaluate the private water supply by tankers. Primary data on water and wastewater quality was acquired by extensive chemical and biological analysis of groundwater samples from a series of wells in the city as well as wastewater samples collected along the effluent channel in addition to flow measurements on the sewage at the wastewater treatment plant. Statistical analysis was performed on several of the collected data sets (i.e. the social economic field surveys, the groundwater quality data, and the irrigation water application) using the software package *SPSS*.

### 1.6.3 Structure of the thesis

The thesis contains two parts, of which the first describes the water use in the city and the second depicts the water use in irrigated agriculture. A final chapter discusses demand as well as supply augmentation emphasizing the wastewater reuse option.

Chapter 2 analyzes characteristics of the domestic water supply in the city of Sana'a, including type and coverage of the water supplies, the water quality of the different supplies, the present and future water demand, the cost and value of water use in the city, and the alternative sources for future water supply. Chapter 3 analyzes the correlation between direct on-site disposal of domestic sewage in the city and the quality of the water supply. This includes examination of the extent of groundwater contamination within the city's boundary and the contamination of the public water distribution network in the eastern part of the city (Nokom) due to sewage logging in the sub-surface soil. Chapter 4 examines the legal and social cultural status of groundwater ownership and use rights in the Sana'a basin from the viewpoint of the government as well as the local farmers. Moreover, it analyzes the implications of the existing institutional setup with regard to resolution mechanisms of water related conflicts. Chapter 5 deals in detail with the agronomic characteristics of irrigated agriculture in the basin area. This involves the analysis of groundwater use for qat and grapes cultivation, including, accessibility to groundwater, irrigation water requirements in relation to actual in-farm water application, and the effects of water availability on land use and cropping pattern. Chapter 6 quantifies the financial incentives for groundwater use for qat and grapes cultivation. More specifically, it examines the cost and profit generation value of groundwater as well as the effects of higher input costs on qat and grapes cultivation, with emphasis on the possible effect of higher energy prices on demand for groundwater. Chapter 7 analyzes the current local wastewater reuse practices. It assesses the available wastewater in terms of both quantity and quality, describes six re-use scenarios, and through the use of multi-criteria analysis it evaluates and classifies the scenarios. In reference to the theoretical framework introduced in Chapter 1, Chapter 8 summarizes the conclusions of the different chapters and identifies the factors that could contribute to the mitigation of the impending water crisis in the Sana'a basin area.

## 1.7 References

Alaerts G. (2000) *Institutional analysis.* Lecture notes, International Institute for Infrastructural, Hydraulic and Environmental Engineering, Delft.

Alaerts G. J. (1996) Generalized analytic framework for institutional studies on water, IHE paper, Delft.

Al-Eryani M., Bamatraf A., Al-Sakkaf G. and Hadash S. (1995) *Socio-economic and Water Rights Aspects.* Technical note No. 17, SAWAS-3, National Water and Sanitation Authority, Sana'a, Yemen.

Al-Garadi M. (1995) The current situation of water production and distribution in the Sana'a city. Proceedings of the extended workshop on the future of Sana'a water supply. National Water and Sanitation Authority, Sana'a, Yemen.

Al-Hamdi M. (1994) Groundwater pollution due to domestic wastewater in the Sana'a city area. Ms.C. thesis no. EE-138, IHE, Delft, The Netherlands.

Al-Hamdi M. (1997) Case study XIII Sana'a, Yemen. In: Helmer R. and Hespanhol I. (eds.). *Water pollution control*, E&FN Spon, London.

Al-Sakkaf R., Zhou Y. and Hall M. (1999) A strategy for controlling groundwater depletion in the Sa'dah plain, Yemen. *Water Resources Development* **15**(3), 349-365.

Bamatraf A. (1995). Irrigation in the Sana'a basin: A challenge facing Sana'a city water supply. Proceedings of the extended workshop on the future of Sana'a water supply. National Water and Sanitation Authority, Sana'a, Yemen.

Beaumont P. (1985) Irrigated agriculture and groundwater mining on the high plains of Texas, USA. *Environmental Conservation* **12**(2), 119-130.

Berkoff J. (1994) *A Strategy for managing water in the Middle East and North Africa.* The World Bank, Washington D.C.

Blomquist W. (1992) *Dividing the Waters-Governing Groundwater in Southern California.* ICS Press, San Francisco, USA.

Caponera D. (1973) *Water Law in Moslem Countries.* Irrigation and Drainage Paper 20/1, FAO, Rome, Italy.

Central Bureau of Statistics (1995) Results of the General Population Census. Ministry of Planning and Development, Sana'a, Yemen.

Charalambous A. (1982) Problems of Groundwater Development in the Sana'a Basin, Yemen Arab Republic. IAHS Publication No. 136, Exeter, England.

Clark C. (1976) *Mathematical Bioeconomics.* New York: Wiley.

Dasgupta P. and heal G. (1979) *Economic Theory and Exhaustible Resources.* Cambridge University Press.

Dinar A., Rosegrant M., Meinzen-Dick R. (1997) Water Allocation Mechanisms, Principles and Examples. Policy Research Working Paper 1779, World Bank.

Dubay L., Thiele J. and Garabedian S. (1984) Sana'a Water Resources, Reappraisal. National Water and Sewerage Authority, Sana'a, Yemen.

Foppen J. (1996) Evaluation of the effects of groundwater use on groundwater availability in the Sana'a basin, Main report. Technical Report No.5, vol. 1, SAWAS-3, National Water and Sanitation Authority, Sana'a, Yemen.

Howard Humphreys (1977) Sana'a Water Supply, phase 2: Interim Report Sana'a Basin. National Water and Sewerage Authority, Sana'a, Yemen.

HWC (1992a) Water Supply, Wastewater, and Sanitation. Final Report Vol. 6, Technical Secretariat of the High Water Council, Sana'a, Yemen.

HWC (1992c) Water Resources management Options in Sana'a Basin. Final Report Vol. 9, Technical Secretariat of the High Water Council, Sana'a, Yemen.

HWC ($1992_b$) Groundwater Resources. Final Report Vol. 4, Technical Secretariat of the High Water Council, Sana'a, Yemen.

Italconsult (1973) Water supply for Sana'a and Hodeida, Sana'a basin groundwater studies. UNDP/WHO, Rome, Italy.

Kahneman D. and Tversky A. (2000) *Choices, Values and Frames*. Cambridge University Press.

Keetelaar C. (1996) Trends in the Water Demand in Sana'a. In: Proceedings of the extended workshop on the future of Sana'a water supply, National Water and Sanitation Authority, Sana'a, Yemen.

Laes D. and Bamatraf A. (1991) Review of the Sana'a Groundwater Monitoring Network. Technical Secretariat of the high Water Council, Sana'a, Yemen.

Laredo D., Haratani J. and McMullan W. (1986) Yemen: Sana'a Basin Water resources Assessment. Wash field report No. 164, U.S. Agency for International Development, Washington D.C.

Libecap G. (1995) The Conditions for Successful Collective Action. In: Keohane, R. and Ostron E. (eds.). *Local Commons and Global Independence, Heterogeneity and Cooperation in Two Domains*. SAGE Publications.

Macoun A. and El-Naser H. (1999) Groundwater Resources Management in Jordan: Policy and Regulatory Issues. In: Salman S. (ed.). *Groundwater, Legal and Policy Perspectives.* The World Bank, Washington D.C.

Ministry of Planning and Development (1998). *Yemen, Human Resources Development*. Sana'a, Yemen.

Moench M. (1993) When Good water Becomes Scarce: Objectives and Criteria. VIKSAT/Pacific Institute Paper.

Moench M. (1994) Approaches to Groundwater management: To control or Enable?. *Economic and Political weekly*. September, 24.

Mosgiprovodkhoz (1986) Sana'a Basin Water Resources Scheme: Summary Report. Ministry of Agriculture and Water Resources, Sana'a, Yemen.

NWSA (1996) Evaluation of the effects of groundwater use on groundwater availability in the Sana'a basin. Technical Report No.5, vol. 5, SAWAS-3, National Water and Sanitation Authority, Sana'a, Yemen.

Ostrom E. (1990) *Governing the Commons: the Evaluation of Institutions of Collective Action*. Cambridge University Press.

Ostrom E., Gardner R. and Waker J. (1994) *Rules, games, and Common-Poll resources*. The University of Michigan Press.

Postel S. (1992) *The Last Oasis-Facing Water Scarcity*. Earthscan Publications Ltd.

Runge C. (1984) Institutions and the Free Rider: The Assurance Problem in Collective Action. *Journal of Politics* **46**, 154-181.

Saleth R. (1996) *Water Institutions in India: Economics, Law, and Policy*. Commonwealth Publishers.

Shah T. (1993) *Groundwater Markets and Irrigation Development: Political Economy and Practical Policy*. Oxford University Press.

Shah T. (1997) Integrating Water Markets in Sustainable Water Resources Management. World Bank Research Observer.

Van der Gun J. and Ahmed A. (1995) *The Water Resources of Yemen: a summary and digest of available information.* Report WRAY-35, Ministry of Oil and Mineral Resources, Sana'a, Yemen.

Wade R. (1988) *Village Republics: Economic Conditions for Collective Action in South Asia.* Cambridge University Press.

Wiesenekker H. (1996) Microbiological Groundwater Contamination of the Sana'a City Area. Technical note No. 23, SAWAS-3, National Water and Sanitation Authority, Sana'a, Yemen.

World Bank (1997) Yemen: Towards a Water Strategy, An agenda for Action. Report No. 15718-YEM, The World Bank.

World Bank (1999) World development indicators. World Bank, Washington D.C. http://www.worldbank.org/data/wdi/home.html.

# Chapter 2

## System and Economic Analysis of Domestic Water Supply in Sana'a

### 2.1 Introduction

With an annual population growth rate of over 8%, Yemen's capital Sana'a is facing an increasing pressure in all of the municipal service sectors. This pressure may even be greater in municipal water supply since the city depends exclusively on a non-renewable groundwater aquifer that is being widely utilized (mined) by agriculture. It is estimated that the water deficit within the Sana'a basin area was around 135 $Mm^3/a$ in 1990 and is projected to reach 400 $Mm^3/a$ by the year 2010 (HWC, 1992). This deficit is currently being satisfied from non-renewable fossil groundwater storage.

Generally, from an economic point of view, the need for water has to be addressed in both quantity and price, and the concept of water scarcity largely reflects the associated costs of the different supply alternatives (Rogers, 1996). In other words, water can be reliably provided for Sana'a if the government or consumers are able and willing to pay the associated costs of desalination and transport[20]. A quick look at the national economy of Yemen suffices to conclude that capacity of consumers to pay the high costs of this water supply alternative is low. Therefore, it becomes quite essential to seek and evaluate other conventional as well as unconventional alternatives in order to utilize the available water resources in the most economic manner, and hence try to mitigate the imminent future water crisis in the area.

Water supply options for the city may be divided into two categories, namely in-basin and out-of-basin alternatives. In-basin alternatives are those resulting from better management of the available groundwater within the basin area, which includes enhanced utilization of rain water, intra-basin transfer and reallocation of groundwater from agriculture to municipal use, especially given the higher value of water in the municipal sector, and utilization of unconventional water resources, namely wastewater reuse. Out-of-basin alternatives include a number of inter-basin transfers of surface water from Surdud, Marib and the Red Sea. These alternatives vary widely in the degree of their technical complexity, costs and social acceptance. Water transfer and reallocation from agriculture to municipal use (either intra-basin or inter-basin) may require the establishment of a political/legal framework that clarifies and defines water use rights and allows for their trade either through land acquisition, direct monetary compensation to the supplying regions, or other compensatory mechanisms.

This chapter aims at analyzing the municipal water supply in the city of Sana'a, from the viewpoints of water managers as well as individual consumers. This chapter will describe and analyze the structure and cost of the domestic water supply market in Sana'a, including: (1) the coverage by the public and private supply systems, their

[20] Sana'a is located at around 200 km from the coast at an elevation of 2,200 meters above sea level.

water quality, and the ensuing choices for consumers, (2) the cost and tariff of the different types of domestic water supply and the consumers' capacity and willingness to pay for water, (3) the present and future domestic water demand, and (4) alternative sources and constraints for future domestic water supply.

## 2.2 Methodology

### 2.2.1 Data on water availability, use and prices of the public and private supplies

Data collection included a wide search for the relevant records, reports and documents from the National Water and Sanitation Authority (NWSA), the National Water Authority (NWRA), the Ministry of Construction, Housing and Urban Planning (MCHUP), the Central Bureau of Statistics and Sana'a University. Interviews were held with officials of the Department of Environmental Health (DEH) of the MCHUP, as well as with various private water suppliers (well owners, tanker owners, and filtration station owners and operators). Data on the costs of drilling and the market prices of pumping equipment were obtained through a survey of some local merchants and contractors.

### 2.2.2 Private non-piped water supply

To analyze financial and economic aspects of private water supply, two field surveys were conducted. The first was a spot-check-survey was conducted at a potable water filtration station. The second survey covered non-potable private tanker water supply. These two supplies have growing coverage and popularity, which is mostly the direct result of limited water supply options.

The objective of the spot-check-survey was to obtain general views and perceptions rather than statistically valid data and based on a fill-and-return questionnaire distributed at *Ghailan* filtration station. Unlike some other stations, where the product water is mainly distributed to supermarkets, the selected station provided an opportunity for direct interaction with consumers since it provides an additional self-filling service. The questionnaire was designed to acquire information on water use, station selection criteria, perceived water quality, tariff and income of respondents. Out of the 60 distributed forms, only 20 replies were received. The lack of a well-defined sample frame introduces uncertainty to the results, limiting generalization.

The second survey was designed to obtain historical, institutional and financial information of tanker water vending. Twelve supplying wells were geographically selected to represent the entire city. From each selling site, 3-5 tanker operators were randomly selected for interviews. In total 47 interviews were carried out. To minimize the interference of the views of one respondent on another, later respondents were selected among those who were not present during the prior interviews. The questionnaires included cross checks especially on questions pertaining to financial matters. Although the sample was randomly selected, the lack of reliable data on the total number of selling sites or tankers makes it difficult to estimate the relation of the sample size to the entire population.

## 2.3 Results and discussions

### 2.3.1 Domestic water supply organization

#### *2.3.1.1 Population*

The population of Sana'a city has been growing rapidly since the early 1970s (Table 2.1), which is attributed to the high national growth rate[21], internal migration from the rural areas and the influx of expatiate Yemenis after the 1990 Gulf Crises.

Table 2.1 Population data of Sana'a city (Central Bureau of Statistics, 1995)

| Date | Population | Annual growth rate |
|---|---|---|
| 1975 National Census | 135,625 | - |
| 1986 National Census | 427,185 | 11* |
| 1994 National Census | 972,011 | 8.83** |

*Based on the two reported figures in 1975 and 1986
**Growth rate reported by the Central bureau of Statistics

As of the end of 1994, Sana'a city constituted 6.15% of the national population, representing 141,938 households with an average of 6.85 persons per household (Central Bureau of Statistics, 1995).

#### *2.3.1.2 The emergence of Sana'a's public water supply*

Prior to the 1970s, domestic water supply in the city of Sana'a depended on water from private open wells distributed by tankers[22], or through small isolated private networks. In 1968, the Ministry of Public Works established a small water supply system, consisting of one drilled well and a small distribution network. After commissioning, management of the system was handed to a local elected user committee. As a result of the continuous expansion of the city, the government established NWSA in 1973 with the responsibility to provide the city with centralized water and sanitation services.

#### *2.3.1.3 Public water supply*

After a major hydrogeological study in the early 1970s and the identification of the Tawilah aquifer as a productive water source with adequate storage to supply the city with domestic water up to the year 2000, a central water supply system was constructed in two phases. The first phase included the drilling of 13 wells in the western wellfield and a distribution network to cover the central part of the city. The second phase of the project was completed in 1982 and included the drilling of an additional 25 wells in the western and eastern well fields and the expansion of the distribution network to reach a design capacity of 400,000 population equivalent. Even though the original master plan included a third phase to cover the entire city with water supply, financial constraints halted the project. Nevertheless, increasing growth in the suburbs of the city, notably Haddah, Nokom, and Al-Rowdda, south, east and north of the city center respectively, pressured NWSA into drilling four wells in each of these areas and supply water through small isolated networks. As of the end of 1994, the coverage of the public water supply was in the order of 45-50%, with 59,100 house connections serving more than 413,000 inhabitants (NWSA, 1994).

[21] As of 1995 the national annual population growth rate was 3.7%.

[22] In the old days, tankers were mostly in the form of 400-liter containers (two barrels welded together), mostly mounted on a carriage pulled by donkeys.

Due to financial constraints and limited availability of water resources compared to demand, NWSA provides intermittent water supply in an effort to maximize its coverage and implement a policy of equitable access to water. For this, the city is divided into two zones, east and west, with each supplied with water every other day. However, the system fails to equally supply every part of each zone due to the systematic low pressure in the system, the limited water availability and the local topography. Therefore, most households have installed ground storage tanks (1-3 $m^3$) to collect water in order to bridge the unreliable 24 hour-cycle supply. The progressive use of larger in-house storage tanks has created problems in the elevated parts of the city, with some areas technically connected, but receiving no water. The low pressure of the public water supply system have forced many consumers to install small pumps to lift water from ground tanks to smaller (1-2 m3) roof tanks in order to provide the necessary pressure head for the house taps. Even though the public water supply is equipped with large storage tanks at elevated locations to balance peak demand and provide pressure in the distribution network, some of these tanks are empty most of the time as a result of limited supply and growing demand. Water consumption is metered, and billed on a monthly basis.

#### *2.3.1.4 Private water supply*

The city of Sana'a was, and still is dependent on private suppliers to satisfy the entire (in the 1960's) or partial (at present) domestic water demand. Whereas in the 1960s private supply often took the form of many small isolated networks connected to open-dug wells equipped with small pumps that ran for a few hours a day, nowadays the predominant form is tanker supplies. Like the public water supply, the intermittent nature of private supplies raises the need for storage tanks. Although piped supplies are perceived by the consumers to be more convenient, financial constraints usually hinder the expansion of private networks confining their coverage to areas around the supplying wells. Despite being a high cost option, tanker supply has the advantage of higher flexibility and reliability for consumers.

Unlike piped supplies, where communication and agreements between the well owner and the various consumers is direct, tanker supply involves a middleman. The tanker owner (middleman) buys water from the well owner at a price that varies according to the tanker size, and subsequently sells it to consumers for a much higher price. The high price levied by tanker owners is argued to cover the high cost of operation and maintenance of the truck and also reflects the long hours of waiting to sell a tanker load. Tanker supply has a greater potential for water conservation due to the higher volumetric tariff structure and leakage reduction, in contrast to the lower flat rate tariff for private piped supplies (Table 2.7).

Like NWSA, all private supplies rely exclusively on groundwater and is mostly used for non-potable uses due to the consumers' perception of its poor (contaminated) quality (Al-Hamdi, 1997) (Chapter 3). Consequently, the separate supply of clean drinking water provided a new commercial opportunity. The first private service started in the early 1990s, with small filtration facilities providing drinking water in 5 and 10 liter containers. The raw water for these stations comes from private tankers or direct connection to private wells. Although these filtration stations are licensed by the DEH of the MCHUP, the interviews revealed major concerns among officials regarding the reuse and cleansing procedures of containers. No technical specifications, other than a "suitable" place and an "appropriate" filtration process, are required for licensing. While the department does not regularly monitor the product water quality, efficiency of the filtration processes is never investigated to validate any real improvements in water quality, thus giving an incentive to overload or bypass the treatment process to increase profits. According to the MCHUP, the number of stations in Sana'a in 1997 was estimated to be around fifty. After cleansing, containers are refilled with filtered water and distributed to supermarkets. In some cases a sticker label of the station is placed across the opening cap. To prevent health hazards associated with the reuse of containers, some consumers have purchased containers that are cleaned at home and refilled at the filtration station on a self-service basis. Water from these filtration stations is mostly used for drinking and to a lesser extent for cooking. Prices of product water vary slightly between stations, but it is common to find double the price at the supermarket (Table 2.7). Even though the results of the spot-check survey at the *Ghailan* filtration station are statistically questionable, they are presented to obtain general views and perceptions of consumers (Table 2.2).

Table 2.2 Characteristics of potable water supply from a filtration station.

| Item | Survey outcome* | Note |
|---|---|---|
| Purchased quantity | 20 l/d | Average |
| Unit cost | 1.5 YR/l (US¢ 1.1/l) | |
| Use | 55% only drinking<br>45% drinking plus cooking | No other use was mentioned |
| Selection criteria | 75% water quality<br>25% convenience | Convenience: close to residence or place of work |
| Previous source for potable water | 70% filtered water from other stations or supermarkets<br>15% Private piped or tanker supply<br>10% NWSA supply | |
| Perception towards water quality | 70% suitable for drinking but does not match bottled water<br>30% suitable for drinking and matches bottled water in quality | |
| Non-potable water source | 90% NWSA<br>10% Private supplies only | The area is covered by NWSA, but irregular supply results in partial reliance on private supplies |
| Perceived NWSA water quality | 85% contaminated, but suitable for non-potable uses<br>15% suitable for drinking | |

* Outcome results pertain to the 20 survey respondents.

### 2.3.2 Water quality

#### *2.3.2.1 Public water supply*

The quality of the public water supply reflects in the first place the groundwater quality at the well fields tapping the Tawilah sandstone aquifer. The main NWSA wells are in the western and eastern well fields located 5-15 km northwest and 5-10 km northeast of the city center, respectively. The water from both well fields is pumped from the wells through collection mains to the main head works at Al-

Hassabah for aeration, chlorination and storage. The raw water is of acceptable drinking quality on all criteria including $NO_3^-$ and coliform bacteria (Table 2.3), hence, chlorination is applied to prevent aftergrowth in the distribution network.

In principle, water quality of each production well is subject to annual monitoring, while monthly quality checks are performed on samples from the main pumping station. Moreover, residual chlorine is tested daily at eight random locations of the distribution network. The chlorine dose is adjusted so as to maintain a residual of 0.1-0.15 mg/l. Water analyses are also carried out when consumers report possible contamination. In cases of confirmed bacteriological contamination, the suspected service area is isolated and disinfected. Despite these checks on water quality, our interviews of consumers revealed a low level of consumer confidence towards the potability of the public supply, in effect limiting its use mostly to non-potable uses.

Table 2.3 Groundwater (wells in the two main well fields) and drinking water quality (NWSA laboratory records; Babaqi and Al-Arossi, 1997)

| Parameter | Unit | NWSA well fields[a] | Private filtration stations[b] | WHO guideline |
|---|---|---|---|---|
| $EC_{25°c}$ | μS/cm | 410-900 | 80-725 | - |
| pH | - | 7-8.4 | 7.7-9.4 | 6.5-8.5 |
| Total alkalinity | mg/l as $CaCO_3$ | 130-300 | 16-136 | - |
| $CO_3^{--}$ | mg/l | 0-15 | 0-24 | - |
| $HCO_3^-$ | mg/l | 100-360 | 20-166 | - |
| Total Hardness | mg/l as $CaCO_3$ | 100-400 | 16-265 | 500 |
| $Ca^{++}$ | mg/l | 75-130 | 5-58 | 200 |
| $Mg^{++}$ | mg/l | 8-40 | 1-36 | - |
| $Cl^-$ | mg/l | 15-100 | 9-125 | 250 |
| $SO_4^{--}$ | mg/l | 30-230 | 0-90 | 400 |
| $NO_3^-$ | mg/l | 0-37 | 4-43 | 45 |
| $Na^+$ | mg/l | 30-110 | 0.7-105 | 200 |
| $K^+$ | mg/l | 2-8 | 0.24-1.7 | - |
| $Fe^{++}$ | mg/l | 0-1.0 | 0.01-0.24 | 0.3 |
| $F^-$ | mg/l | 0-1.3 | 0.04-1.65 | 1.5 |

a Available records of NWSA wells in the two main well fields(1982-1993).
b Concentration range of 8 tested stations.

Babaqi and Al-Aroosi (1997) found that tap water from the public distribution network contained high counts of both coliform and *E.coli.* However, they did not report the exact location of the sampling point, and it can not be excluded that this may be attributed to bacteriological contamination near the household storage tank or the internal house plumbing.

### *2.3.2.2 Private water supply*

The water supply from private sources (piped, tanker and container supply from filtration stations) is not subject to regular governmental quality control. Several studies have identified many private wells to be contaminated with nitrate and coliform bacteria, thus confirming pollution from on-site domestic wastewater disposal (Chapter 3). Many of the tested private wells either contained high coliform counts, nitrate concentration exceeding the recommended WHO drinking guidelines, or both. An investigation of the water quality of eight filtration stations (Table 2.3), and the two major water bottling companies, revealed that while the tested bottling companies comply with the WHO drinking water guidelines, six treatment stations had bacteriological contamination which was suspected to be the result of contaminated raw water in combination with ineffective filtration, or, more probably, due to unhygienic handling and refilling of containers.

### 2.3.3 Unaccounted-for-water

#### *2.3.3.1 General*

Three components of unaccounted-for water can be distinguished by physical leakage from the distribution network, illegal connections and unpaid service. In other words, unaccounted-for water is the difference between produced and sold quantities. Physical leakage comprises major bursts of supplying mains and small more or less continuous leaks from secondary branches. Although water loss from the first category is high in a given small time scale, easy and quick detection[23] and repair limit the loss of considerable quantities of water over longer periods of time. Detection of the second category is, however, more difficult, and thus may lead to large cumulative water losses. Corrosion, faulty joints, deteriorated service connections or faulty valves can result in perforated pipes and small leaks. Illegal connections can be identified through a systematic inspection, and in the case of metered systems by a careful analysis of the data from consumer and distribution meters (UNDTCD, 1992).

#### *2.3.3.2 Public water supply*

Prior to 1992, water production from the NWSA well fields was not metered and the only source of data was that obtained from consumer meter readings. As a first approximation, sold quantities were taken to constitute 80% of total production while the remaining 20% was considered to represent an "acceptable" unaccounted-for water. Metering of all production wells in 1992 lead to a more precise assessment of the total production quantities, revealing a 41% unaccounted-for water. Subsequent measures in 1993 and 1994 reduced the unaccounted-for water to 38% and 36%, respectively. These high percentages are suspected to be the result of: (1) water looting at the wells by the local guards[24] to irrigate their land, (2) illegal connections and unpaid service, and (3) physical leakage from faulty joints, fittings, and bursts of PVC pipes installed in the first construction phase of the system. A study aimed at investigating the performance of the distribution network concluded that water looting between the two main well fields and the main headwork is in the order of 20% of the total water production, or 50% of the total unaccounted-for water (NWSA, 1992; Haidera, 1995). It was estimated that tighter control to eliminate water looting could reduce the unaccounted-for water from 40% to 20% in 1992, which could free over 3.9 $Mm^3/a$ of water production to expand the coverage in the city and supply 130,000 additional consumers, or improve the service to current customers.

#### *2.3.3.3 Private water supply*

Private piped water supplies in Sana'a can be assumed to have a very low unaccounted-for water. Low pressure, small size of the coverage area, short supply time, and quick response to major breakage, limit physical leakage to negligible

[23] Water discharge from the ground, loss in pressure, or soil subsidence.

[24] Since no compensation was handed for confiscated well sites, original land owners were employed as guards as a way of indirect compensation.

quantities. It is also expected that illegal connections are quite rare in private systems due to the small and limited boundaries of the coverage area where any illegal work near the network is easily detected. Immediate cut-off of water supply usually follows unacceptable delay in payments, thus eliminating any supply to an unpaid connection. As for tanker supplies, except for minimal spillage during tankers' filling, unaccounted-for water does not exist. Unaccounted for water in the case of filtration stations is also minimal and is limited to container cleansing.

#### *2.3.3.4 Structure of domestic water demand*

Two variables influence the domestic water demand, namely population and per capita water consumption. To reach a realistic estimate of the overall domestic water demand for Sana'a, the coverage and the per capita consumption rates of public and private supplies are quantified (Table 2.4).

Table 2.4 Coverage of the different suppliers in domestic water use in Sana'a (1995).

| Supply Type | | Value |
|---|---|---|
| (a) Public supply | | |
| Total water production (NWSA records) | (a) | 50,685-52,055 $m^3$/d |
| Unaccounted for water (estimated 35%) | (b) | 17,740-18,220 $m^3$/d |
| Water consumption | (c)=(a)-(b) | 32,945-33,835 $m^3$/d |
| Non-domestic use (estimated 5%) | (d) | 1,650-1690 $m^3$/d |
| Domestic consumption | (e)=(c)-(d) | 31,295-32,145 $m^3$/d |
| Reported house connections | (f) | 59,100 |
| Total population | (g) | 972,011 inh. |
| Population density per household | (h) | 7 inh./household |
| Served population | (I)=(f)*(h) | 413,700 inh. |
| Per capita water consumption | (e)/(i) | 75-78 l/d |
| Service coverage | (j)=(i)/(g)*100 | 43% |
| Service coverage (incl. 2-7% unpaid, or illegal connections) | | 45-50% |
| (b) Private supplies | | |
| Per capita water consumption[25] (Assumed) | (k) | 40 l/d |
| Service coverage | (l) | 50-55% |
| Total private water consumption | m=l*k*g/1000 | 19,440-21,380 $m^3$/d |
| Total domestic water consumption | (n)=(e)+(m) | 50,735-53,525 $m^3$/d |
| Total domestic water production | (a)+(n) | 70,125-73,435 $m^3$/d |

In order to forecast the domestic water demand for 2010 and 2025 (Table 2.5), the following assumptions are made:

- Future coverage level of both public and private supplies remain constant at the 1995 levels.
- Per capita water use of the public supply drops slightly to 70 l/d (based on NWSA estimates).
- leakage drops to 20%.
- Per capita water use of the private supplies remains at the current estimate of 40 l/d.

[25] The low per capita water use is justified by the higher price of private supplies than that of the public supply.

- Demand projections are based on three scenarios of annual population growth (low 3.7%; medium 5%; and high 9%). Moreover, 1995 is used as the base year for future population projections.

Table 2.5 Projections of future domestic water demand for Sana'a (all demand in $Mm^3/a$).

| Item | Low growth rate (3.7%) | | Moderate growth rate (5%) | | High growth rate (9%) | |
|---|---|---|---|---|---|---|
| | 2010 | 2025 | 2010 | 2025 | 2010 | 2025 |
| Population (Inhabitants) | 1,676,300 | 2,890,900 | 2,020,700 | 4,200,900 | 3,540,500 | 12,896,300 |
| Demand on Public system | 24.1-26.8 | 41.5-46.2 | 29.0-32.3 | 60.4-67.1 | 50.9-56.5 | 185.3-206.0 |
| demand on private systems | 12.2-13.5 | 21.1-23.2 | 14.8-16.2 | 30.7-33.7 | 25.8-28.4 | 94.1-103.6 |
| Total demand | 36.3-40.3 | 62.6-69.4 | 43.8-48.5 | 91.1-100.8 | 76.7-84.9 | 279.4-309.6 |

Given the relaxed governmental policy towards population and urbanization, it is unlikely that the growth rate of the city will fall to the national average of 3.7%. The low employment opportunity in the rural areas suggests that internal migration to the urban areas will continue in the future. It is therefore quite appropriate in the scope of this discussion to assume that the moderate scenario growth of 5% is most likely. Therefore, total domestic water demand is due to increase to around 45 and 95 $Mm^3/a$ by the year 2010 and 2025, respectively.

### 2.3.4 Cost and value of domestic water supply in Sana'a

#### *2.3.4.1 Cost of groundwater pumping*

It is assumed that the tariff of the various types of domestic water supply does not stand in direct relation to the production costs, particularly in the case of private supplies. Given that groundwater is the exclusive source of domestic water supply, the associated capital and recurrent costs of water pumping are examined (Table 2.6). While the capital costs were obtained from interviews with merchants and contractors, yield estimates and the operation and maintenance costs were obtained from a survey of 70 agricultural wells within the Sana'a basin (Chapter 5). All prices are given so as to reflect the construction and operation of a typical borehole with a total depth of 300 meter and a water level at 120 meter below ground surface.

Although the calculated average unit cost (Table 2.6) does not include any treatment or distribution components, it is still more than twice the first block of the public water supply tariff.

Table 2.6 Typical pumping cost of deep groundwater in Sana'a (all prices in US$; 1998 average parallel market exchange rate of 1 US$ to 135 YR).

| Item | Cost (US$) |
|---|---|
| (a) Capital costs | |
| Drilling (45-60 US$/m)[a] | 13,500-18,000 |
| Diesel generator | 10,000-15,000 |
| Submersible pump | 5,500-6,500 |
| Pipes (100-110 US$ per 3-m length to a depth of 135 m) | 4,500-5,000 |
| Cables (15-18 US$ per m to a depth of 140 m) | 2,100-2,500 |
| Motion converter | 1,500-2,000 |
| Pump house | 500 |
| Total capital costs . | 37,600-49,500 |
| Amortized cost based on a lifetime of 20 years and interest rate of 7% (US$/a) | 3,550-4,780 |
| Costs to offset an annual 3-m drop in water level[b] (US$/a) | 250 |
| Total annual capital cost | 3,800-4,930 |
| Unit cost:<br>(Based on an estimated average well yield of (7 l/s) and 12 operating hours per day, results in annual abstraction of 110,000 $m^3$). | 3.5-4.5 US¢/$m^3$<br>(4.7-6.1 YR/$m^3$) |
| (b) Recurrent costs | |
| Fuel (diesel + oil):<br>(based on average value (chapter 6)) | 4.9 US¢/$m^3$ |
| Maintenance:<br>(based on average value (chapter 6)) | 3.7 US¢/$m^3$ |
| Recurrent unit cost | 8.6 US¢/$m^3$<br>(11.6 YR/$m^3$) |
| Total unit cost | 12.1-13.1 US¢/$m^3$<br>(16.3-17.7YR/$m^3$) |

a Reported value is for private contracts, but may increase to as much as 150 US$/m for official (governmental) contracts.
b This includes capital (a 3-m pipe, cable, etc.) and installation costs.

### 2.3.4.2 *Tariff of the public water supply*

NWSA has adopted a progressive block tariff rate since its establishment in 1973 in order to encourage water conservation. Over the past twenty years the tariff has developed as can be seen from Table 2.7. Even though the increasing block tariff rate conflicts with economies of scale, where the unit cost decreases with higher production, in many cities of developing countries this system is often employed to encourage water conservation and in some instances promote cross subsidies. Despite the growing water scarcity in Sana'a, the tariff of the public water supply is considered very low and is for example only around half of that in Jakarta, Indonesia[26] and Rabat, Morroco (Lovei and Whittington, 1993; McPhail, 1993), where a typical household (7 inhabitants and 70 l/cap.d) pays a monthly tariff of 1.02, 1.94 and 1.91 US$ in Sana'a, Jakarta and Rabat, respectively.

[26] The water tariff for Sana'a is US$ 0.05/$m^3$ for the first 10 $m^3$, US$ 0.11/$m^3$ for the second 10 $m^3$, US$ 0.15/$m^3$ for the third 10 $m^3$, US$ 0.22/$m^3$ for the fourth 10 $m^3$, and US$ 0.30 for any consumption higher than 40 $m^3$ per month. On the other hand, the water tariff in Jakarta is US$ 0.1, 0.2, 0.3 and 0.5 for the first 15 $m^3$, second 15 $m^3$, third 20 $m^3$ and beyond 50 $m^3$ per month, respectively.

Table 2.7 Development of water tariff of the public water supply in Sana'a.

| Period | Tariff | | Comments |
|---|---|---|---|
| | YR/m$^3$ | US$/m$^3$ | |
| Prior to 1979 | 2 | 0.44 | Tariff for first block of 10 m$^3$. |
| 1979-1981 | 10 | 2.22 | Tariff for first block of 5 m$^3$. |
| 1981-1985 | 3.6 | 0.80 | Tariff for first block of 7 m$^3$. |
| 1985-1996 | 5.4 | 1.2-0.045 | Tariff for first block of 10 m$^3$. Change in tariff in US$ is the result of market devaluation of the local currency from 4.5 YR for 1 US$ in 1985 to around 120 YR for 1 US$ in 1996. |
| 1996-present | 7 | 0.05 | Tariff for first block of 10 m$^3$. 1998 average Parallel exchange rate is 135 YR for 1 US$. |

Table 2.7 shows that though the tariff has more than tripled since 1979, taking the currency devaluation into account, it has actually decreased by a factor of 8.8 (not counting any inflation on US$). Despite the progressive tariff adopted by NWSA, a household that consumes the excessive amount of 27m$^3$ per month[27] is charged only YR360 (2.7 US$), which is still cheaper than a private 3 m$^3$ tanker load of non-potable water.

The erosion of the real tariff has lead to the following consequences:

(1) For those who live close to the main pumping station or along the distribution primary mains, water is readily available and there is no deterrent for water wastage.

(2) The current tariff structure results in a lack of proper cost recovery, which in turn influences the following financial and organizational aspects:

(a) All capital costs are irrecoverable and thus must be subsidized.

(b) The collected tariff is inadequate to perform proper operation, preventive maintenance, etc., resulting in deteriorating service level.

(c) Renovation and expansion of the system depend exclusively on the availability of subsidized funds from internal (government) or external (donors) sources.

It may therefore be concluded that while some low-income households may benefit from the current tariff, the real beneficiaries of this policy are the larger water consuming middle and high-income households. The deteriorating infrastructure and service level will lead to higher renovation and expansion costs that will have to be borne by the next generation. Moreover, over half of the city's population is also affected by the current tariff policy of the public water supply since they rely on the more expensive private supplies creating a non-equitable access to water.

#### *2.3.4.3 Tariff of private water supplies*

Private water supply alternatives present a wide tariff range (Table 2.8). An important cause of this variation is their unregulated nature with water vendors setting their prices according to personal and local circumstances attempting to maximize profit.

[27] With 7 persons per household, this would correspond to a per capita consumption of around 130 l/d, comparable to that in some industrialized countries, and around twice the actual average consumption.

Table 2.8 Tariff of private water supplies in Sana'a (1998) (field data).

| Service type | Type of use | Tariff type | Tariff / Price | |
|---|---|---|---|---|
| | | | **YR** | **US$*** |
| Private (piped) | Domestic non-potable | Un-metered monthly fixed rate per household.<br>Approx. per $m^3$ | 500-900<br>40-70 | 2.22-6.67<br>0.3-0.52 |
| Private borehole | Domestic non-potable | Price to tankers per $m^3$ | 33 | 0.25 |
| Private tankers | Domestic non-potable | Delivered to household | 160-200 | 1.2-1.5 |
| Filtered water from filtration stations | Mostly drinking | Price per $m^3$ | 1500-2000 | 11.1-14.8 |
| Filtered water from supermarket | Mostly drinking | Price per $m^3$ | 4000 | 29.6 |
| Bottled water | Drinking (only) | Price per $m^3$ | 26,667-33,000 | 197-244 |

* Prices in US$ based on the parallel exchange rate of 135 YR.

The tariff structure of both the public and private water supplies (Tables 2.7 and 2.8) show the large price range for water, which may be the result of growing water demand, limited supply, and the lack of a regulatory framework. Similarly, water tariff in Jakarta, Indonesia, shows a high variation ranging from 0.1-0.5 US$/$m^3$ for house connections to the public water supply, 1.8 US$/$m^3$ for tanker supply and 1.5-5.2 US$/$m^3$ for service through distribution vendors (Lovei and Whittington, 1993).

*2.3.4.4 Economics of domestic water supply in Sana'a*

In order to calculate the economics of water use for the public water supply, various data, which are mostly unavailable, are needed. Therefore, the following assumptions are used:

- Even though the capital costs for the NWSA wells are higher than these presented in Table 2.6 due to the higher construction standards and the installation of higher capacity pumps, the unit cost of the capital investment may be assumed comparable or even lower than those presented in Table 2.6 due to scale effects[28]. Therefore, a capital cost component of 4.7-6.1 YR/$m^3$ is used.
- Although no information is available regarding the capital cost of the distribution network, it is assumed that the unit cost component of the capital investment for the distribution system is in the order of 2-3 times that of water production (drilling), resulting in a unit cost of 9-18 YR/$m^3$.
- The 1994 NWSA records show an operation unit cost of 4.55 YR/$m^3$. Although preventive maintenance of the NWSA supply is neglected due to lack of funds, it may be assumed at around 30-50% of the operational costs. Therefore, a total unit cost of 6.0-6.9 YR/$m^3$ is estimated for O&M.

Using the above estimates, Table 2.9 shows the economics of the public water supply.

[28] On average a NWSA well yields around 4.5 times the average yield of a private well (NWSA, 1994).

Table 2.9 Economic status of the public water supply in Sana'a

| Capital costs ($YR/m^3$) | | O&M costs | Total costs | Tariff | Economic |
|---|---|---|---|---|---|
| **Drilling** | **Distribution** | **($YR/m^3$)** | **($YR/m^3$)** | **($YR/m^3$)** | **setting** |
| 4.7-6.1 | 9.0-18.0 | 6.0-6.9 | 19.7-31.0 | 15.7* | (Deficit) 4.0-15.3 $YR/m^3$ |

* This is the composite tariff of the first two blocks based on a water consumption of 14.7 $m^3$/month for a typical household, and incorporates a connection fee of 11,700 YR, which translates into a unit price of 6.2 $YR/m^3$.

Although Table 2.8 show the selling price for the varies forms of private water supplies, it is essential to examine other costs in order to analyze the economics of water supply to consumers (Table 2.10).

Table 2.10 Income generation value of water for private domestic suppliers.

| Supply type | Cost type | Cost ($YR/m^3$) | Price ($YR/m^3$) | Profits ($YR/m^3$) | Return (%) | Annual net income[a] (1000 YR) |
|---|---|---|---|---|---|---|
| On-site (well) Direct selling to tankers. | Capital | 4.7-6.1 | 33 | 15.3-16.7 | 86-100 | 1,396-1,524 |
| | Recurrent | 11.6 | | | | |
| | Total | 16.3-17.7 | | | | |
| Piped supply House connection from a private well | Capital (well) | 4.7-6.1 | 40-70[d] | 16-37 | 50-155 | 242-560[d,e] |
| | Capital (transport)[b] | 4.7-9.2 | | | | |
| | Recurrent (Well) | 11.6 | | | | |
| | Recurrent (transport)[c] | 2.9-5.8 | | | | |
| | Total | 24-33 | | | | |
| Tanker supply[f] | water | 33 | 166-200 | 60-94 (37-71)[f] | 57-90 (29-55)[f] | 214-335 (132-253)[f] |
| | O&M of truck | 73 | | | | |
| | Total | 106 (129)[f] | | | | |

a Based on the results of the survey on tanker water supply (see section 3.5.4).
b Based on the assumption that capital transport costs are 1-1.5 times the capital costs of pumping.
c Based on the assumption that recurrent transport costs are 0.25-0.5 times that of pumping.
d This is based on a moderate per capita water use of 60 l/d.
e Based on an assumed 100 household piped connections.
f Capital cost of tankers was neglected since the survey indicated that almost all trucks have practically exceeded their economic lifetime. Nevertheless, if a capital cost component is added, using the average price of 600,000 YR for a tanker (from the survey), an extended life of 10 years and an interest rate of 7%, the unit cost of water is adjusted as seen above.

All figures may be converted to US$ by using the 97/98 average parallel exchange rate of 135 YR/US$.

Table 2.9 shows that the deficit of the public water supply is partially being subsidized[29] and would gradually result in lower service level and limited system life,

[29] Using the 1995 NWSA water production records, it can be calculated that the real annual subsidies were in the order of YR 74-291 million.

due to poor maintenance and the lack of cost recovery. The costs and tariffs of the different water supply options in Sana'a are presented in Figure 2.1.

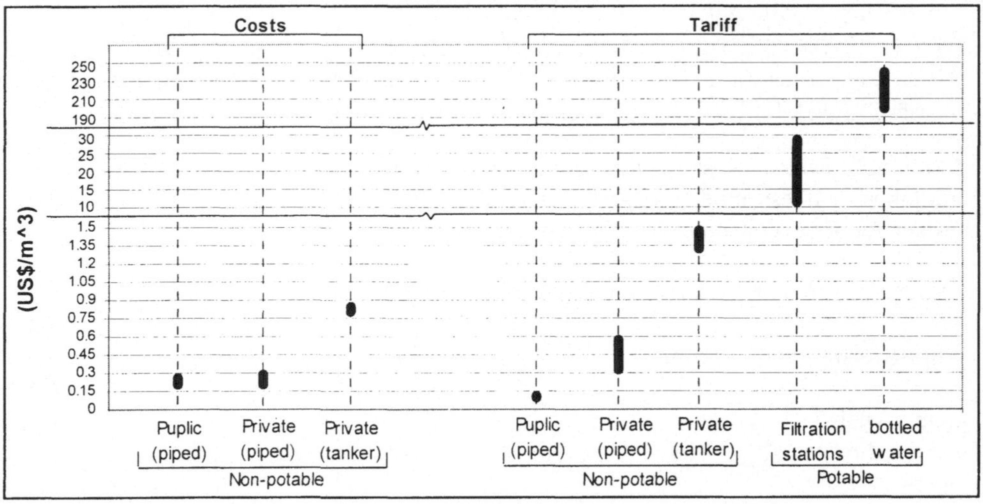


Figure 2.1 Cost and tariff of the different water supply alternatives.

### *2.3.4.5 Cost and tariff of sanitation*

Sana'a was partly sewered in the mid 1980s and at present the sewerage system covers around 20-25% of the city's population (chapter 7). The tariff of the public sewerage system is incorporated in the water bill and uses the same progressive block rate system. The remaining larger portion of the population relies on on-site cesspits for sewage disposal. Table 2.11 shows the tariff structure of the sewerage system and the unit cost of on-site disposal facilities.

Table 2.11 Tariff and cost of wastewater disposal.

| **Service type** | **Tariff of sewerage system and unit cost of on-site cesspits** |
|---|---|
| Sewerage system | 10.0 YR/m$^{3}$ [a] (7.4 US¢/m$^{3}$) |
| On-site cesspit | 27-47 YR/m$^{3}$ [b] (20-35 US¢/m$^{3}$) |

a This is the composite tariff of the first two blocks based on a water consumption of 14.7 m$^{3}$/month for a typical household, and incorporates a connection fee of 6,000 YR, which translates into a unit price of 3.2 YR/m$^{3}$ (2.4 US¢).

b This is based on the following assumptions:

- A total capital cost of 50,000 YR (370 US$) for a 20m deep cesspit.
- Amortized cost of 4,720 YR (35 US$) based on a lifetime of 20 years and 7% interest rate.
- Daily water consumption range of 40-70 l/cap, and an average of 7 inh. per household.

The low tariff of the sewerage system reflects a lack of proper cost recovery, thus halting any effort for sustained expansion and renovation of the collection and treatment systems until external funds are secured.

### *2.3.4.6 Value of water to consumers*

The different assumptions used to calculate the share of water and sanitation in the overall budget of a typical household in Sana'a are listed below:

- A total household monthly income range of 20,000-50,000 YR.
- Per capita water use of 70 l/d for those connected to the public supply.

- A monthly flat tariff range of 500-900 YR per household for those relying on piped private supply.
- Per capita water use of 40 l/d for those relying on private tanker supplies.
- A purchase of 5 l/d per household for drinking from filtration stations.

With the exception of the first, all other assumptions are based on collected data, either from the field or from documented reports. It is therefore essential to validate the income range before it is used to drive the relative portion of the income used for water supply and sanitation.

The income range was selected based on the National Household Budget Survey, carried out by the Central Bureau of Statistics in 1998 (Ministry of Planning and Development, 1998a-d). The survey results were presented in four consecutive reports. Although the final report is not out yet, Table 2.12 shows the income level for urban households as they appear in the available preliminary reports. Although the average income is quite high, the selected income range is justified by the lower median of the income, indicating a large portion of the population with lower than the reported average income levels.

Table 2.12 Average and median income of urban households in Yemen (Ministry of Planning and Development, 1999a-d)

| **Report** | **Average income (YR/month)** | **Median income(YR/month)** |
|---|---|---|
| 1st quarter report | 37,848 | 25,000 |
| 2nd quarter report | 38,089 | 27,000 |
| 3rd quarter report | 36,083 | 25,000 |
| 4th quarter report | 35,925 | 25,000 |

Using the monthly income range of 20,000-50,000 YR, Table 2.13 shows the relative share of water and sanitation on the overall budget of the household based on the type of water and sanitation services.

Table 2.13 Share of water supply and wastewater disposal in the budget of a typical household in Sana'a.

| **Type of service** | | **Share of drinking water (%)** | **Share of non-drinking water (%)** | **Share of Sanitation (%)** | **Total (%)** |
|---|---|---|---|---|---|
| **Water supply** | **Sanitation** | | | | |
| Public (piped) | Sewerage | 0.6-3 | 0.5-1.2 | 0.3-0.7 | 1.4-4.9 |
| | On-site | | | 0.7-3.1 | 1.8-7.3 |
| Private (piped) | Sewerage | 0.6-3 | 1.0-4.4 | 0.2-0.6 | 1.8-8.0 |
| | On-site | | | 0.6-2.7 | 2.2-10.1 |
| Private (tanker) | Sewerage | 0.6-3 | 2.7-8.4 | 0.2-0.4 | 3.5-11.8 |
| | On-site | | | 0.4-1.8 | 3.7-13.2 |

Unlike many cities in developing countries, where the tariff of private water supplies changes according to the season (rainy vs. dry), the tariff in Sana'a seems to be insensitive to this factor which is mainly attributed to the low rainfall (200-250 mm/a), higher water consumption during the summer (rainy) months and the high cost of rainfall harvesting by the individual households. As for the percentage of income spent on water supply, some studies have shown that while poor households may spend as much as 18% (Onitsha, Nigeria), 20% (Port-au-prince, Haiti), and 9% (Addis Ababa, Ethiopia), middle income households may spend 5% (Morogoro, Tanzania and 9% (Ukunda, Kenya), and the upper income households spend only around 2-3% (Onitsha, Nigeria) of the total income on water (Linn, 1983; Whittington et al., 1989; Katko, 1991; Whittington et al., 1991). MaPhail (1993) reports that residents of Rabat's shantytowns, Morroco, were willing to pay 2% of their estimated income, which is the same tariff rate levied on dwellers of Rabat city, for house connections despite the existing free, reliable and perceived good quality standpipe service.

Based on the service type and income level, Table 2.13 provides an approximation of the share of water supply and sanitation services in the overall household income in Sana'a. A household with a monthly income of 50,000 YR obtaining potable water from filtration stations, connected to the public distribution network for non-potable supply, and connected to the sewerage system is expected to spend only around 1.4% of the income on water and sanitation. On the other hand, a household with a monthly income of 20,000 YR obtaining potable filtered water from supermarkets[30], relying on tankers for non-potable water, and depending on a cesspit for wastewater disposal is expected to spend the higher budget of 13.2% of the income on water and sanitation.

Table 2.13 shows also that the separate purchase of drinking water by those relying on public water supply absorbs a higher share in the household budget than for non-potable water, which results from the substantially cheaper and under-priced non-potable bulk public water supply. Consumers of private piped, and to a larger extent, tanker supplies, on the other hand, are faced with higher prices of non-potable bulk supply, thus lowering the relative share of drinking water in the overall budget. It can thus be argued that, faced with a low share of water in the household budget, consumers of the public water supply would prefer high, reliable water quality for drinking and thus would be willing to pay the higher prices for the perceived better water quality from the filtration stations. On the other hand, faced with more expensive non-potable water, poor consumers of private supplies may prefer to make an effort to reduce the share of water in their overall budget and, hence, may not value enough the higher-quality water, leading to possible use of the cheaper questionable quality bulk water for drinking as well.

The survey on tanker supply shows that 16% of tanker supply consumers are rich, while 30% and 54% are of poor and middle income, respectively. These data however, are based on information from the tankers owners and operators, and thus may be subjective since no precise criteria could be used to identify each income class. This information contradicts the expectation that rich households rely more on the expensive tanker supply since the public water supply physically restricts their higher demand, and they tend to live in newly urbanized areas not covered by NWSA. Nevertheless, these results could highlight the social consequences of the irregular NWSA service, forcing many poor households to also depend on tanker supply.

The survey results confirm that tanker water supply is a relatively a new and growing service. Eighty nine percent of the respondents started water vending during the last decade (1988-1998) and 40% of those started only during the last three years (1986-

[30] Filtered water are also sold at super markets for as much as 2.5 times the price at the filtration station.

1998). Only 5% of the respondents started commercial water vending prior to 1985. This trend in water supply is a strong evidence of the rapid growth and increasing reliance on private supplies. While at present all tankers obtain water from wells within the city's perimeter, it is expected that as water becomes more scarce, tanker owners would start to fetch water from wells in the neighboring rural areas thus incurring higher costs that will be reflected in higher prices to consumers.

The survey results depict tanker water vending in Sana'a as a small-scale individual business, where 94% of surveyed tanker operators are also owners. Selling sites of water through tankers are usually situated at the same location of the supplying wells. According to the survey results, decisions of tanker owners to select a selling site, and thus the supplying well, are based on two main criteria, namely water quality and water demand. Although water quality was the first default answer of most respondents to justify the selection of the selling site, it is believed that this justification is only used as a promotional argument with higher water demand being the real reason. At most selling sites tanker owners have to register in order to able to sell at that particular location. Registration at some sites involve paying a small fee paid to an elected person to supervise the selling activities according to a first-filled first-sold principle. Although there is no seasonal variation in water price, the per tanker average daily number of sold loads in summer and winter are 4.2 and 2.0, respectively. In the surveyed sites, the average number of tankers per selling site is 24 with variation between 5 and 50. Using the average capacity of 3 $m^3$ per load, the average sold water quantity is 9.75 $m^3$/d/tanker and 250 $m^3$/d/selling site.

### 2.3.5 Water resources

#### *2.3.5.1 Current water resources*

In the past, several spring-fed brooks crossed Sana'a. Water was bowled from these brooks and from many open wells to satisfy household water demand. It is most likely that these springs were fed by the local groundwater system. With the continuous decline in groundwater levels, these springs eventually disappeared. Therefore, the local surface water in the past was only part of the local complex groundwater system. Climatic conditions with a low rainfall and high evaporation potential in the Sana'a area highly reduce the chance for permanent availability of surface water.

Groundwater has always been an important water source in Sana'a. The Tawilah sandstone is the most important aquifer within the Sana'a basin has been heavily exploited since the early 1970s. As of 1990, groundwater abstraction was estimated at 180 $Mm^3$/a, of which 83% (150 $Mm^3$/a) is estimated to have been withdrawn for irrigation (laredo et. al, 1986; HWC, 1992). The heavy reliance on groundwater has resulted in substantial drop (2-6 m/a) in water levels (Laes and Bamatraf, 1991). Estimates of the natural recharge to the basin vary between 28-60 $Mm^3$/a, supplemented by the infiltration of domestic sewage (13-15 $Mm^3$/a), and the return flows from irrigation.

#### *2.3.5.2 Future water resources*

Faced with rapid depletion in groundwater storage, the Sources for Sana'a Water Supply (SAWAS) project was initiated in the mid-1980s, with the objective to identify and evaluate new water supply sources for Sana'a. The project identified seven feasible alternatives, which range from further development of the Tawilah aquifer to and transport of desalinated seawater from the coastal area (Table 2.14).

Table 2.14 Alternative water supply sources for Sana'a (Kruseman, 1997).

| Alternative | Estimated yield ($Mm^3/a$) | Resource type |
|---|---|---|
| Sabaeen Park well field | 10.4 | Development of Tawilah aquifer south of Sana'a. |
| Sana'a South well field | 3.2 | Development of volcanic aquifer south of Sana'a. |
| Well field east of Shibam | 9.5 | Development of Tawilah aquifer east of Sana'a. |
| Kharid Dam | 7.9 | Development of surface water from the Kharid spring, north of Sana'a. |
| Wadi Surdud Reservoir | 15.8 | Development of surface water from Wadi Surdud, west of Sana'a. |
| Marib Dam Reservoir | 15.8 | Development of surface water from The Marib lake, east of Sana'a. |
| Red Sea desalination | 15.8 | Desalination of seawater and piped transport. |

While the first three alternatives are just further mining of the same mostly non-renewable groundwater resource, the next two are development of surface water sources with suspected connection to groundwater aquifers. Surface water from the Marib reservoir and desalinized seawater from the Red Sea seem to be the only independent reliable long-term options. Although some alternatives have the potential to provide more water than others, their costs and yield vary widely (Figure 2.2).

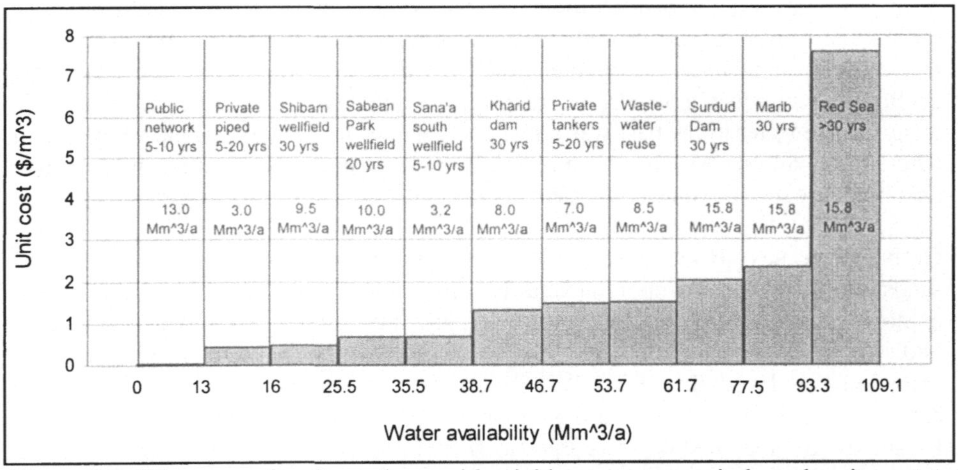


Figure 2.2 Water supply alternatives with yield, coverage period, and unit cost to consumers (some values adapted from SAWAS, 1997).

Figure 2.2 is a supply curve showing the yield of the alternative sources, their cost, and expected time coverage. The option of wastewater reuse is included as a potential source[31], where the estimated unit cost is based on reported costs of some existing reclamation projects (Chapter 7).

Although it is quite difficult to estimate or project the price elasticity of water in Sana'a, Figure 2.2 suggests that the unit cost of the first 7 supply alternatives fall within the current price range of water, and, thus may be considered within the current capacity and willingness of consumers to pay. With regard to the option of wastewater reclamation, even though the estimated unit cost is comparable to that for private tanker supply, public reaction and acceptance to reuse reclaimed wastewater for domestic uses cannot be fully verified (Chapter 7). The last three alternatives, with a unit cost range of 2-7.5 US$, are probably beyond the capacity of the public to pay for water and suggest the need for high subsidies, especially if the last alternative is to be implemented.

*2.3.5.3 Constraints to water supply augmentation*

Developing and importing surface water from Surdud, Marib, and the Red sea face high costs that range from 1.7 to 6.4 times higher than the current expensive option of private tanker supply. In addition to the high costs, major political, social, legal and technical constraints to water transfer are expected (Figure 2.3).

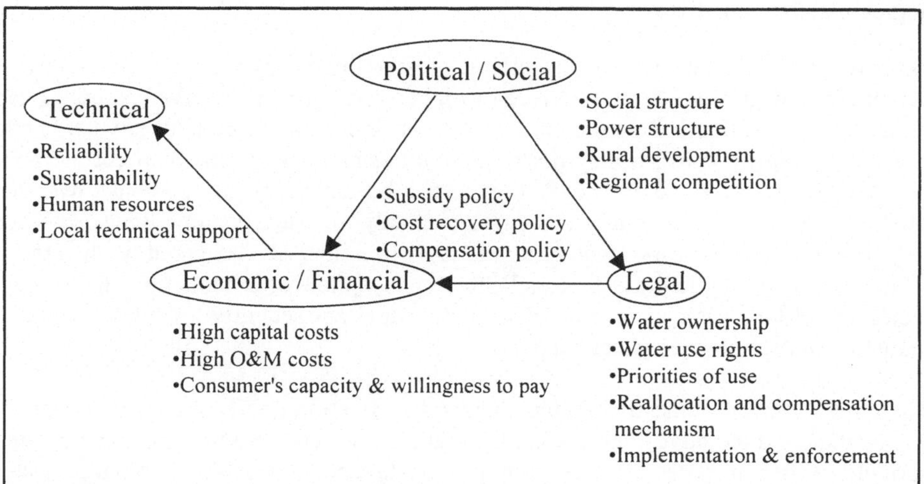


Figure 2.3 Constraints of water import to Sana'a

Importing water from Marib, and to a lesser extent from Surdud, will face major resistance from current water users (mostly farmers) in these regions. The general governmental view towards ownership of water resources[32] as well as the religious and customary principles of priorities of water use[33], make such water transfers and reallocation theoretically feasible. However, the local social and power structure and a

[31] The potential yield from wastewater reuse (8.5 $Mm^3/a$) is based on the assumption of an expansion of the sewerage system to cover 20% of the city in 2010.

[32] The constitution of the country states that all natural resources are state property.

[33] Religious and customary water use principles give first priority to human potable use followed by livestock drinking and then irrigation.

low development level in the resource regions[34] complicate implementation of any reallocation plan. Issues of water use rights are likely to be settled if the rights of the local users are defined and acknowledged. The introduction of a community based compensation strategy for such customary user rights (through purchase or trade), could stimulate the creation of a suitable environment for water transfers, especially if the compensation funds are used for the development of the supplying regions through local elected committees.

A second important obstacle for water reallocation is the prohibitive cost associated with treatment and transport of water from the supplying regions. In addition to financial compensations to the supplying regions, the large investment of these schemes will exert a high financial burden on the government and consumers who will have to bear at least part of the costs, which will exceed their willingness and capacity to pay. Almost all of these supply schemes involve a high level of technical difficulty that includes pumping and transfer over long distances and advanced treatment. These technical constraints become of greater importance in Yemen due to the limited local industrial capacity and scarce skilled human resources.

Although the above constraints seem to be prohibitive, the acute water shortage in the area faces an imminent implementation of at least some of the mentioned augmentation alternatives.

Globaly, demand management is gaining greater importance in the area of water resources management. While the role of higher water prices of private supplies can be viewed as an effective instrument to lower water consumption, with low per capita water consumption rates, it may not be desirable since it will mostly affect the poor resulting in a lower water use that may not be sufficient for personal hygiene. Moreover, due to the generally low price elasticity of water, water saving from use reduction in the domestic sector is relatively minimal if compared with similar reductions in the agriculture sector. With the exception of a reduction in the high unaccounted-for water, it seems that there little applicability of water demand reduction in the domestic water supply of Sana'a.

Therefore, since domestic water demand reduction and inter-basin supply augmentation alternatives have clear limitations, water reduction in the irrigated agricultural sector seems to be more promising and could result in large water savings. The major part of groundwater from the Tawilah aquifer is currently used for irrigated agriculture. Since cereals have been driven out of cultivation, irrigated agriculture within the Sana'a Basin is in the form of cash crops, namely qat, grapes and vegetables. High profits from cash crop cultivation stimulate over-irrigation, thus wasting large quantities of valuable water. Subsidized energy (electricity and diesel) for agricultural production (mainly groundwater pumping) contributes to unrealistically low water cost and thus inefficient irrigation practices. As argued in chapters 5 and 6, better management of water use in irrigated agriculture can free large quantities of water for more "valuable" uses. Assuming a 30% saving, through technical and agronomic demand reduction measures to improve irrigation efficiency, around 50 $Mm^3/a$ could be freed to be reallocated for domestic use, thus delaying the need for expensive water transfer schemes. However, It should be realized that the

[34] Farmers in these areas argue that their regions have been neglected in the national development plans.

first step to address the problem of water shortage in Snaa'a, is to prevent any further expansion in groundwater use for agriculture.

## 2.4 Conclusions and Recommendations

The coverage of the public and private domestic water supplies are around 45-50% and 50-55%, respectively. Assuming moderate growth (5%) and similar to present water use (40-70 l/cap.d), future water demand for Sana'a is expected to increase to 45 and 95 $Mm^3/a$ for the year 2010 and 2025, respectively. Groundwater from the NWSA well fields are of acceptable drinking quality, but may deteriorate during transport in the distribution system. Private non-potable and drinking supplies are of questionable quality due to the contamination potential of the supplying wells and unhygienic handling of containers in filtration stations.

At least a real reduction by a factor of 8.8 has taken place in the tariff of the NWSA supply since the late 1970s. With low cost recovery potential and the rapid depletion of the local groundwater resources, NWSA is constrained to maintain its coverage and invest in new expensive supply schemes. The total share of water (potable, non-potable and sanitation services) in the overall budget of a typical household ranges from 1.5-13.5%, depending on the type of service and income.

Future water supply augmentation alternatives include water import from Surdud, Marib and desalinized water from the coastal areas. Water transfer from these regions is expected to face major social, political, financial and technical constraints. Demand management measures are not expected to result in large water savings in the domestic water supply sector as a result of the already low water use. Given the large-scale groundwater use in irrigated agriculture, demand management may lead to significant water savings that can be reallocated for domestic use thus delaying the need for expensive water transfer schemes.

## 2.5 References

Babaqi A.and Al-Aroosi M. (1997) The quality of the bottled and container treated water in the city of Sana'a. Faculty of Engineering, Sana'a University, Sana'a, Yemen.

Central Bureau of Statistics (1995) Results of the General Population Census. Ministry of Planning and Development, Sana'a, Yemen.

Haidera M. (1995) Computer modeling of water distribution systems with intermittent supply, case study: Sana'a (Yemen). Ms.C. thesis no. EE-194, IHE, Delft, The Netherlands.

HWC (1992) Water Resources management Options in Sana'a Basin. Final Report Vol. 9, Technical Secretariat of the High Water Council, Sana'a, Yemen.

Katko T. (1991) Reselling and vending water. *Journal of the American Water Works Association JAWWA5,* 83(6), 63-69.

Kruseman G. (1997). Exclusive Summary. Sources for Sana'a Water Supply (SAWAS) project, National Water and Sanitation Authority, Sana'a, Yemen.

Laes D. and Bamatraf A. (1991) Review of the Sana'a Groundwater Monitoring Network. Technical Secretariat of the high Water Council, Sana'a, Yemen.

Laredo D., Haratani J. and McMullan W. (1986) Yemen Sana'a Basin Water resources Assessment. Wash field report No. 164, U.S. Agency for International Development, Washington D.C.

Linn J. (1983) *Cities in the developing world.* New York: Oxford University Press.

Lovei L. and Whittington, D. (1993) Rent-extracting behavior by multiple agents in the provision of municipal water supply: a study of Jakarta, Indonesia. *Water Resources research*, 29(7), 1965-1974.

McPhail, A. (1993) Overlooked market for water connections in Rabat's Shantytowns. *Journal of Water Resources Planing and Management*, **119**(3), 388-404.

Ministry of Planning and Development (1999a) Results of the first round of the Household Budget Survey (1998) (January-March), Sana'a, Yemen.

Ministry of Planning and Development (1999b) Results of the second round of the Household Budget Survey (1998) (April-June) Sana'a, Yemen.

Ministry of Planning and Development (1999c), Results of the third round of the Household Budget Survey (1998) (July-September), Sana'a, Yemen.

Ministry of Planning and Development (1999d) Results of the fourth round of the Household Budget Survey (1998) (October-December), Sana'a, Yemen.

NWSA (1983-1994) Un-compiled laboratory records. Central laboratory, National Water and Sanitation Authority, Sana'a, Yemen.

NWSA (1992) Annual report, 1992 for Sana'a branch, National Water and Sanitation Authority, Sana'a, Yemen.

NWSA (1994) Annual report, 1994 for Sana'a branch, National Water and Sanitation Authority, Sana'a, Yemen.

Rogers, P. (1996) Water: The economic and financial Context. A paper presented at a regional consultation meeting, Towards a policy for water resources development and management in the Asian and Pacific regions, The Asian Development Bank, Manila, The Philippines.

UNDTCD (1992) Demand management. International Conference on Water and the Environment, Dublin.

Whittington D., Lauria D. and Mu X. (1991) A study of water vending and willingness to pay for water in Onitsha, Nigeria. *World Development*, **19**(2-3), 179-198.

Whittington D., Lauria D., Okun D. and Mu X. (1989) Water vending activities in developing countries: A case study of Ukunda, Kenya. *International Journal of Water Resources Development*, **5**(3), 158-168.

# Chapter 3

# Contamination of Water Supply in Sana'a

## 3.1 Introduction

Provision of safe drinking water is not only considered a service, but in many countries is also viewed as an effective policy to safeguard the public health from water borne and water related diseases. The connection between some diseases and contaminated potable water is very well established and the outbreaks of epidemics caused by contaminated water supply systems in many countries are evidence to such connection.

One of the primary pollution sources of water supply systems is domestic wastewater. In many cases, urban wastewater is collected by sewerage systems and transported to treatment plants, where the degree of treatment depends on the adopted standards or regulations. The treated effluent is finally disposed off either to the sea (some coastal cities), to surface water, or reused for agriculture, industrial, or domestic purposes.

A second disposal method, widely used in developing countries, is on-site disposal of raw sewage either to open drains, or to the underground through leaching cesspits or septic tanks followed by infiltration pits. Disposal of raw sewage to open drains leads often to the contamination of the receiving water body thus reducing its usefulness as a possible water supply source, or increasing the risk of disease outbreaks in case of illegal/unregulated use and consumption of such water source. As for wastewater disposal to the underground, the situation is different in three aspects: (1) the relatively higher self purification capacity of the soil and underground layers, (2) the larger time scale for any pollution to travel, and (3) the very high cost of groundwater rehabilitation. Since groundwater is often of very high and clear potable quality, allowing it to become contaminated carries a disproportionally high opportunity cost, especially in water scarce areas that depend exclusively on groundwater for domestic supply.

Drinking Water in distribution networks too can be affected by sewage through cross-connections between water supply mains and sewerage lines, or through back-siphonage (partial vacuum). Back pressure in the distribution network may result in the infiltration of some wastewater through breakage, or faulty fittings and connections of the water supply pipes. This may occur when sewage is allowed to percolate to the underground in areas with low soil infiltration capacity causing wastewater ponding and saturation of the top soil cover[35] with wastewater. To minimize the contamination risk, the network pipes should be selected as to resist the corrosive nature of wastewater and adequate pressure should be applied continuously in the entire network to prevent any suction of wastewater from the surrounding environment. The contamination risk of the distribution network is greatest when

[35] Water distribution pipes are usually laid at this depth.

wastewater ponding and pressure fluctuation in the network occur simultaneously. Although such cross-contamination is assumed to occur by all handbooks and reports on water supply in developing countries, it has not actually been measured to take place under controlled experimental circumstances.

Due to the rapid growth of the city of Sana'a, and as a result of limited financial resources, the National Water and Sanitation Authority (NWSA) in recent years has not been able to maintain nor expand its water supply and sanitation coverage. At present, the coverage of the public water supply is in the order of 40-50%, and only 15-25% for sewerage (chapter 2). Over half of the city's population relies on unregulated private water vendors, who in turn fetch water from unmonitored private wells scattered within the city's perimeter. From the health and environmental viewpoints this is alarming since around 75% of the population, or over 700,000 inhabitants, discharge their domestic wastewater to on-site cesspits, thus threatening to contaminate the same aquifers used for domestic water supply. Groundwater in the Sana'a area is the exclusive water source, and is mostly non-renewable[36] (fossil). Reliance on groundwater for both urban domestic use and irrigated agriculture in the surrounding rural areas has resulted in heavy mining leading to an annual drop of 3-6 m in groundwater level since the early 1980s, reflecting a growing serious water scarcity in the area.

Not only does widespread on-site disposal of raw sewage in Sana'a actually lower the quality of the local groundwater aquifers, but it also threatens to contaminate the NWSA distribution network in areas prone to clogging of cesspits, and wastewater ponding. The combination of wastewater ponding and pressure fluctuations in the distribution network exist in al least two quarters of the city, namely Nokom and Al-Lakamah, east and west of the city center, respectively (Figure 3.1).

This chapter will analyze the consequences of direct disposal of raw domestic sewage on the quality of water supply in Sana'a. Therefore, this chapter will (1) examine the overall extent of deep aquifer contamination within the city's boundary for the period 1993-1997 and attempt to relate such contamination to factors like population density and sewerage coverage, and (2) investigate the contamination of the NWSA distribution network in the Nokom, quarter of Sana'a, by subsurface contamination. Before the specific condition of water supply contamination in Sana'a is discussed, section 3.2 introduces some general examples and experiences of water supply contamination by domestic wastewater, discusses public health implication, and estimates the total wastewater production in Sana'a.

## 3.2 Contamination of water supply

### 3.2.1 Groundwater contamination from domestic wastewater disposal

Contamination of groundwater from domestic wastewater takes place through widespread use of on-site disposal facilities, leakage from sewer lines or infiltration from treatment facilities such as oxidation ponds.

In the Gaza strip, Palestine, as of 1995, the number of cesspits was estimated at 100,000 to cover around 90-95% of the population. They discharge around 30-40

[36] The annual natural recharge in 1995, for example, is estimated to have constituted only 10-30 % of the total abstraction (Foppen, 1996).

$Mm^3/a$ of domestic sewage to the subsurface environment. This has resulted in groundwater contamination with nitrate concentrations exceeding 100 mg/l $NO_3$-N in several locations (Al-Agha, 1995; Nashashibi and Van Duijl, 1995).

In response to the growing scarcity of freshwater, Egypt has initiated large-scale wastewater reclamation projects for agricultural use in desert areas. In an assessment of the impact of a wastewater reclamation project on the quality of groundwater in the area of Gabal El-Asfer, 20 km northeast of Cairo, it was concluded that long-term wastewater irrigation has led to a decrease in the salinity of the originally brackish groundwater. It was also concluded that groundwater was enriched with nitrogen (ammonium and nitrate), phosphorus, heavy metals and coliform bacteria (Farid et al., 1993; Rashed et al., 1995).

Another typical example of groundwater contamination by on-site disposal of raw sewage, is the situation in Dar-es-Salam, Tanzania. Around 60% of the population of the city, depends on a combination of septic tanks and soak pits for sewage disposal resulting in elevated levels of ammonium and fecal coliform concentrations in the upper unconfined shallow aquifer (Gondwe et al., 1997). Until recently, the city of Tehran, Iran, depended mainly on on-site seepage pits, pit-privies, and leaching cesspools for municipal and industrial wastewater disposal. In a survey of groundwater quality, virtually all zones of the city showed microbial contamination of groundwater throughout the year (Shariatpanahi and Anderson, 1987). In Chennai (Madras), India, sewage from leaking sewers, soakaways, and septic tanks, and river infiltration, have been identified as probable sources of significant microbiological and nitrate contamination of groundwater. Nitrate concentrations in some parts of the city were as high as 1000 mg/l, while 70% of the samples exceeded the Indian drinking water standard of 45 mg/l. This is especially alarming as a large proportion of the city's population relies on shallow wells for water supply.

In an investigation of groundwater quality in the city of Coventry, UK, concentrations of nitrate, chloride, sulfate, calcium, magnesium, and metals were found to be higher in groundwater samples taken from the urban area in comparison to the reference groundwater quality in the surrounding rural areas. It was concluded that the change in groundwater quality should be attributed to spillage, deliberate disposal, and leaking sewers and drains (Lerner et al., 1992). Groundwater contamination with domestic wastewater in Germany has been mainly attributed to leaking sewerage systems, which would require an investment of around 50-100 billion DM (30-60 US$ billion) to correct (Bucksteeg, 1991). As of 1980, over a third of houses in the state of Virginia, USA, used some type of on-site wastewater disposal system, representing 238,000 $m^3/d$ of domestic wastewater from 10% of the urban and nearly 75% of the rural houses. The combination of high leakage density and poor soils suggests that on-site sanitation facilities contribute substantially to groundwater contamination. Despite this, the trend in using on-site wastewater disposal was growing (IJzerman et al., 1992). Nitrogen in groundwater in central Long Island, New York, USA, has been attributed to lawn and garden fertilizer in addition to wastewater disposal from cesspools and septic tanks (William et al., 1984). Florida, USA, counts an estimated 600-700 wastewater disposal (injection) wells in addition to around 25,000 septic tanks. Fecal indicator bacteria have been detected in groundwater and surface water near areas served by on-site disposal facilities. A poliovirus was isolated from a 30.5 meter deep well, where the source was identified to be a septic tank drain. It was concluded that the virus passed through 5.5 m of clay, 2.5 m of shale and 22.5 m of limestone to reach the well. In Long Island, New York, enteric viruses were isolated at distances of up to 67 m from the leaching beds of a subsurface wastewater disposal system (Paul et al., 1997).

In an experiment, viruses seeded in a septic tank were detected in groundwater three days to nine weeks after inoculation. Moreover, the appearance of viruses correlated with periods of increased precipitation, pH, and total coliform concentrations. Viruses were observed in groundwater for over 100 days after leaving the septic tank (Rao et al., 1987). In an earlier investigation in England, enteric viruses were regularly isolated from groundwater that contained zero count of indicator bacteria. This confirms that viruses have a longer survival time in the subsurface environment and are able to travel longer distances, than coliform bacteria, which raises the question whether fecal coliform bacteria are the best biological contamination indicators (Slade, 1981).

The transport of organic substances in the soil and their infiltration/penetration mechanisms into groundwater are mostly unclear. Results of a study in Israel indicate that anionic and non-ionic detergents are found in municipal wastewater of several coastal cities. It was suggested that at least 20% of the non-ionic surfactants incorporated into detergent formulations in Israel reach surface or groundwater sources. Furthermore, it was also concluded that groundwater is being contaminated by non-ionic surfactants whose origin are surface streams polluted with municipal sewage, especially given the fact that the currently employed wastewater treatment processes cannot guarantee total removal of detergents (Zoller, 1993).

### 3.2.2 Pollution indicators

The selected pollution indicator is usually related to the type of discharged waste. In the case of groundwater contamination by domestic wastewater, the typical contamination indicators include notably fecal coliforms, nitrogen (both as ammonium and nitrate), chloride, BOD (or COD) and total dissolved solids. In case of on-site disposal of sewage, biological metabolization in the septic tank or cesspit, and later within the soil and subsoil layers, usually consume the available organic matter thus gradually reducing both BOD and COD to low levels. Depending on the extent and type of chemicals used in homes (disinfectants, detergents, etc.), chemical compounds too may find their way to the groundwater. Nevertheless, some organic toxicants are usually undetectable due to the limitation in detection technology at low concentrations. In general, the strong correlation of nitrogen concentration, fecal coliform count and the total dissolved solids with contamination of water supply, together with their simple and affordable detection, make them suitable indicators of domestic sewage pollution.

### 3.2.3 Health concerns of groundwater and drinking water contamination

Health effects caused by the contamination of drinking water by domestic wastewater can include outbreaks of water borne diseases usually associated with gastrointestinal symptoms, and increased carcinogenicity created by certain organic chemicals. Bacterial pathogens known to have occurred in contaminated drinking water include *Salmonella*, *Shigella*, enterotoxigenic *Escherichia coli*, *Vibrio cholerae*, *Yersinia enterocolitica*, and *Campylobacter*. These organisms cause diseases, which vary from mild gastroenteritis to severe and sometimes fatal dysentery, cholera or typhoid. However, the infectious dose required to cause human infection varies from one organism to another (Feachem et al., 1977). The significance of the water route in the spread of intestinal bacterial infections varies considerably both with disease type and the local circumstances. Excessive nitrate concentrations in drinking water (>45 mg $NO_3^-$/l) pose a health hazard especially to infants. Moreover, cancer may result from nitrosamines, related to high nitrate or nitrite concentrations (William et al., 1984).

In Sweden, 63 waterborne disease outbreaks were reported between 1980 and 1989 resulting in a total of 32,250 reported illness cases (Anderson, 1991). In Finland, there were 24 waterborne disease outbreaks resulting in 10,000 reported cases between 1980 and 1992. Sixty percent of the reported outbreaks in Finland, representing 75% of the illness cases, were attributed to contaminated groundwater from public as well as private wells. Almost 17% of the epidemics were the result of direct contamination of groundwater by domestic sewage from leakage or blockage of sewers in the vicinity of water supply wells. A cross-connection between a drinking water supply main and a sewer was also identified as the cause of at least one outbreak (Lahti and Hiisvirta, 1995).

The small size of viruses (20-85 nm) gives them a high mobility in groundwater. Even though adsorption of viruses to the soil particles is considered to be an effective removal mechanism, under certain conditions viruses have been detected several hundred meters away from the point source. Viruses are considered to have caused some 12% of the total number of waterborne disease outbreaks in the United States between 1946 and 1980. Nevertheless, difficulties in isolating many virus types lead to an underestimate of the reported figure. The average number of documented waterborne disease outbreaks in the United States has increased by almost four-fold since the 1960s, which is attributed to higher reporting and enhanced surveillance techniques. Hepatitis A virus, Horwalk virus, and rotavirus have been identified as causes of groundwater disease outbreaks in the United States (Gerba and Rose, 1990). In an earlier study, 20% of 99 samples from shallow groundwater wells tested positive for enteric viruses (Marzouk et al., 1979). Gerba (1987) also showed that almost 20% of samples collected from the southeastern United States were positive for enteric viruses. In both studies, the tested wells were used as a source of non-chlorinated water supply to communities and households. Viruses have also been detected in treated water supply that passes the drinking standards for coliforms.

Between 1971 and 1982, there were 399 waterborne outbreaks resulting in 86,050 reported cases of illnesses in the United States. The share of groundwater in those outbreaks is shown in Table 3.1.

Table 3.1 Share of groundwater related waterborne outbreaks in the USA 1971-1982 (total of 399 outbreaks and 86,050 illness cases) (after Craun, 1985).

| | **Disease outbreaks Percentage of total/(No.)** | **Illness cases Percentage/(No.)** |
|---|---|---|
| Untreated groundwater (well water only) | 28% / (110) | 10% / (8,558) |
| Inadequately disinfected groundwater (well water only) | 17% / (67) | 25% / (21,488) |

Of all water borne disease outbreaks connected to the use of untreated groundwater in Table 3.1, the overflow or seepage of sewage through soil, limestone, and fractured rocks was responsible for almost 43% of the reported 110 groundwater related outbreaks (Craun, 1985). Between 1981 and 1988, consumption of untreated or inadequately disinfected groundwater caused 44% of the reported 248 waterborne outbreaks in the United States. Most outbreaks and illnesses attributed to distribution networks are due to cross-connections and contamination of distribution storage. It was concluded that 13% of the total reported outbreaks between 1981 and 1988 in the United States could be attributed to contamination of distribution networks resulting from cross-connections and repair of mains. It was also concluded that chlorine application in small amounts to prevent bacteriological after-growth may not be adequate in cases of source contamination (Craun, 1991).

Fifty three percent of the reported cases of virus presence was related to water receiving complete treatment, which indicates that viruses are more resistant to removal and inactivation than bacteria. It has been reported that some viruses survive 1.0 mg/l of chlorine for 15 minutes of contact time, which questions again the adequacy of using coliform bacteria as a base for biological drinking water standards (Gerba and Rose, 1990)

Even though many of the waterborne disease cases are not reported, the figures cited above on contamination of groundwater and drinking water are alarming. This is even of more importance in developing countries where water borne and water related diseases seem to be even more common. Although official records of water borne diseases are not available due to poor surveillance and reporting, in Yemen the regular media coverage of diarrheal disease events reflects the magnitude of the problem.

Unlike water borne diseases that show immediate consequences, long term consumption of water with low concentrations of particular organic or inorganic chemical compounds may cause chronic diseases, notably cancer. Some cancer-causing chemical compounds are usually present in wastewater or result from the oxidation of organic matter during treatment.

### 3.2.4 Domestic wastewater disposal in Sana'a

The limited water use in the past influenced both domestic wastewater quantities and disposal methods. While dry disposal of excreta was widely practiced, limited gray water[37] was usually directed to local gardens for irrigation. As water use increased, water-based disposal methods became a widespread practice leading to greater reliance on on-site cesspits. In the early 1980s, a sewerage system was constructed in Sana'a to cover the old part of the city in addition to areas along the collection mains to a temporary treatment plant consisting of a series of oxidation ponds. Since construction, the sewerage system has been expanding (mainly through more connections and not physical expansion), which resulted in overloading conditions of the oxidation ponds and an increasing percentage of bypassed flows of raw sewage. Collected wastewater from Sana'a was measured in 1994 and 1997 to average 10,200 and 14,200 $m^3/d$, respectively (chapter 7). Based on the calculated domestic water consumption (Chapter 2), the coverage of the sewerage system and the on-site disposal of domestic wastewater in Sana'a in 1995 can be estimated as follows:

[37] Wastewater from kitchen activities in addition to that from washing, bathing and laundry.

(a) Sewerage system

| | |
|---|---|
| Total domestic water consumption: | 50,700-53,500 $m^3/d$ |
| Assume 90% wastewater disposal[38]: | 45,600-48,200 $m^3/d$ |
| Collected wastewater: | 10,200 $m^3/d$ (3.7 $Mm^3/a$) |
| Coverage of sewerage system: | 20% of population |

(b) On-site disposal

| | |
|---|---|
| Thus, coverage of on-site disposal: | 80% |
| Raw sewage to cesspits: | 35,400-38,000 $m^3/d$ (13-14 $Mm^3/a$) |

## 3.3 Approach and methodology

Sampling programs were carried out on (1) groundwater from the different zones of the city and (2) the distribution network at the area of Nokom. Chemical and biological analysis of pollution indicators is used to confirm contamination. Statistical analysis of subsequent sampling rounds is used to elucidate the spatial and temporal variability of groundwater contamination.

### 3.3.1 Groundwater sampling

From an inventory of around 140 private wells (NWSA, 1996), 29 boreholes were selected for sampling (W1 through W29) (Figure 3.1). In selecting the boreholes, depth and type of use were the main selection criteria so as to represent the deeper aquifer used for domestic supply. Therefore, the selected wells range between 150 to 350 meters in depth and are used for a variety of domestic uses. Given the general direction of groundwater flow in the area (south to north), the wells W1-W9 represent the southern part of the city, while W10-W25 and W26-W29 represent the middle and northern parts, respectively. For the majority of the boreholes, samples were collected during continuous pump operation; for the rest, operators were asked to start the pumps and let the water run for a few minutes prior to sampling[39]. Samples were collected between 8:00 and 10:30 a.m., and then transported to the NWSA central laboratory for analysis.

After confirmation of groundwater contamination from the initial sampling session in November 1993, it was decided to conduct another session of groundwater analysis of the selected wells. Consequently, a second sampling session in September 1997 was carried out. It should be noted that due to logistical constraints it was not possible to sample all 29 wells during the second sampling session (Table 3.2).

Table 3.2 Non-sampled wells during the different sessions.

| Sampling session | Non-sampled wells |
|---|---|
| 1993 | None |
| 1997 | W6, W8, W12, W14, W25 |

[38] 10% of water use is assumed to be lost through consumption, evaporation, spillage, etc.
[39] This was done to ensure freshness of the samples.

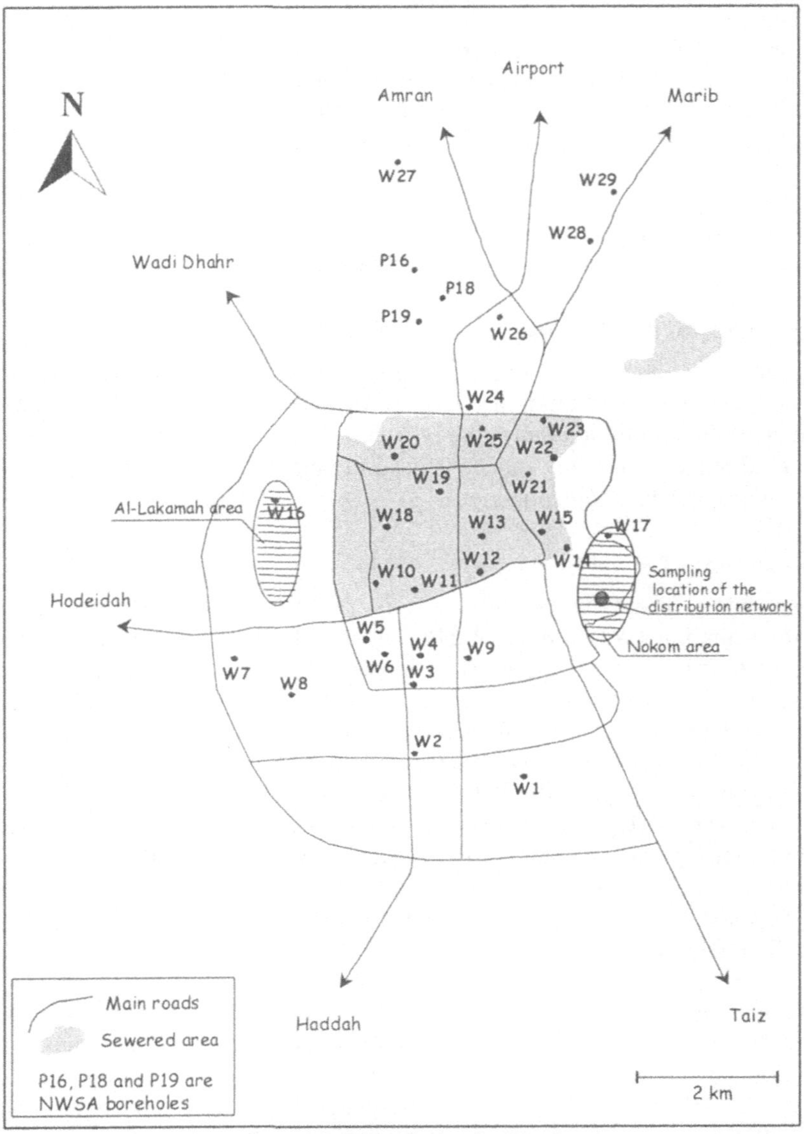


Figure 3.1 Location of sampled boreholes in Sana'a.

### 3.3.2 Sampling from the public distribution network in Nokom

In order to confirm the relationship between a change in the water quality within the distribution network and the presence of wastewater in the immediate vicinity of the water supply pipe, sampling would have to take place so as to collect any traces of sucked-in wastewater during a previous sudden drop to negative pressure, e.g., immediately after closure of the main valve. It is believed that any sucked-in wastewater would be flushed out of the network when water is supplied again 24 hours later, therefore it was decided to collect a sample from a suitable site within the network immediately after valve opening (and pressure build-up) so as to collect any traces of contamination. Collecting further samples at small time intervals should show a declining trend in the concentration of indicators to within background range, as any sucked-in wastewater gets fully flushed from the network. By comparing the quality of the first few samples to the background water quality of the supplying source, contamination can be confirmed.

The area of Nokom is located in the eastern part of the city (Figure 3.1). The topography of the area is steep since it is located at the lower slopes of the Nokom Mountain. An isolated distribution network, which is not connected to the city's main network, supplies this area with water. The network is fed from three wells owned by NWSA. The wells are equipped with submersible pumps with an average production of 1340 $m^3$/d, 1460 $m^3$/d, and 1820 $m^3$/d for M1, M2 and M4, respectively. The pumps are connected directly to the network, where they produce a pressure head of up to around 60 meter water column (mwc) in the lower part of the distribution network. The supplied area is divided into two parts, with each part being supplied with water every other day, thus inducing large pressure fluctuations in the network.

The Nokom area is not connected to the sewerage system and relies entirely on cesspits for wastewater disposal. However, the area is most probably located on an impermeable rock formation that inhibits infiltration, leading to rapid clogging of cesspits and consequent ponding and overflow of sewage onto the streets. The prevailing situation with its high fluctuation in water pressure and the presence of wastewater near the distribution mains makes this area particularly suited for this investigation.

Three sampling sessions were conducted in December '93, and for results confirmation another two sessions during September '95. These sessions will be referred to as Sessions 1, 2 and 3 to represent the 1993 sampling sessions and Sessions 4 and 5 to represent the 1995 sampling sessions. Disturbances encountered during Sessions 1, 2 and 4 are listed in Table 3.3. During Sessions 3 and 5, arrangements[40] were taken to eliminate any external disturbances. With the exception of Session 2, the results of all chemical and biological analyses are discussed in section 3.5.2. Session 2 was considered insufficiently representative.

Table 3.3 Disturbances encountered during sampling.

| Session No. | Disturbance | Complication |
|---|---|---|
| 1 | Proportions of water supplied by each well could not be determined. | Could not verify whether the elevated indicator concentrations in the samples were the result of contamination during distribution, or due to the different concentrations of the supplying wells. |
| 1 | High coliform counts lead to the identification of a break in the distribution main. | The results could not be generalized to circumstances pertaining to a "normal" network. |
| 2 | Power was cut off to one of the supplying wells minutes prior to sampling. | Sudden changes in the quality at the sampling point. |
| 4 | Some of the main valves were not tightly closed. | Suspected discharge of any contamination from leaky valves during non-supply cycle. |

[40] Arrangements included: The repair of the pipe breakage followed by a chlorine disinfection in that part of the network. Only one supplying well was in operation prior to and during the sampling day, thus background water quality could be easily identifiable. A check to identify and repair leaky valves in the area was carried out.

The samples were collected at the moment of opening the main valve (around 8:00 am) to supply that particular zone and thereafter at small time intervals (Table 3.4). They were collected from a private house connection located about 10 meters away from the main water supply pipe and around 70 meters away from the main controlling valve. In order to eliminate any contamination from the household storage tank or improper internal plumbing, samples were collected from a point before the house water meter. After sampling, all samples were immediately taken to the laboratory for analysis.

Table 3.4 Time intervals for sampling from the public water distribution network in Nokom, Sana'a (minutes). T=0.0 is moment of valve opening.

| **Session** | **Sample** | | | | | | | | | |
|---|---|---|---|---|---|---|---|---|---|---|
| | (1) | (2) | (3) | (4) | (5) | (6) | (7) | (8) | (9) | (10) |
| (1) | 0.5 | +1.0 | +1.5 | +2.0 | +3.0 | +4.0 | +5.0 | +6.0 | +7.0 | - |
| (3) | 0.0 | +1.0 | +2.0 | +3.0 | +4.0 | +5.0 | +10.0 | +15.0 | +20.0 | +25.0 |
| (4) | 0.0 | +0.5 | +1.0 | +1.5 | +2.0 | +2.5 | +3.0 | +4.0 | +5.0 | +10.0 |
| (5) | 0.0 | +0.5 | +1.0 | +1.5 | +2.0 | +2.5 | +3.0 | +4.0 | +5.0 | +10.0 |

### 3.3.3 Indicator parameters

Given the composition of wastewater from Sana'a (Chapter 7), the typical domestic pollution indicators $NO_3^-$, $NH_4^+$, $Cl^-$ and the electrical conductivity (EC) were selected to be measured. Low ammonium concentrations were found in the '93 groundwater sampling session, which could be attributed to nitrification and/or adsorption to the soil particles. This eliminated the need for ammonium measurements during the '97 groundwater sampling sessions. The same indicator parameters, in addition to both total and fecal coliforms, were also selected for the analysis of samples from the distribution network. Background concentrations of unpolluted groundwater and of the water source of the distribution network are used for comparison purposes.

### 3.3.4 Chemical and biological analysis

All samples were analyzed at the central NWSA laboratory, where chemical and biological analysis were conducted according to standard procedures (APHA, 1980). The membrane filtration method was used for total and fecal coliform tests. The HACH DR/2000 Direct Reading Spectrophotometer was used to measure the concentrations of both $NH_4^+$ and $NO_3^-$. Duplicate measurements were conducted of each sample. EC was measured using the calibrated Wissenschaflich-Technische Werkstatten (WTW) EC meter. Based on the sample's temperature, a factor was used to correct the measured value as to represent the $EC_{standard}$ in μS/cm at 25 °C.

### 3.3.5 Statistical analysis

The statistical software package *SPSS* was used to analyze spatial and temporal variation of $NO_3^-$ and EC concentrations in the collected groundwater samples. Three statistical tests were performed as shown in Table 3.5. As for the first test, wells W1-W9 were designated to represent the southern zone of the city, while W10-W25 and W26-W29 were set to represent the central and northern zones respectively. Nitrate concentration in each zone was compared to those of the other two zones, resulting in a total of three comparisons for each of the '93 and '97 sampling sessions. The same procedure was also conducted for the EC values.

The second statistical test was intended to analyze the relationship between the dependent variable (in this case the $NO_3^-$ concentration or EC) and the independent variables of population density and sewerage coverage within the city. The rationale is to use the independent variables, to predict groundwater pollution represented by $NO_3^-$concentrations, or EC values, thus relating groundwater contamination to the intensity of water use or wastewater disposal. To overcome the lack of population density data, based on the author's knowledge of the neighborhood, a scale of 0-10 was used to represent the approximate number of houses surrounding each well to a radius of around 40-50 meters. Using this procedure, the wells were proportionally scaled according to the population density. Another local person was asked to scale the wells using the same criteria, and the average of the two values was used in the test. As for the sewerage coverage, the wells were given a score of 1 or 0 based on whether the area of the well is sewered or not.

Table 3.5 Statistical tests used to evaluate the change in $NO_3^-$ concentration and EC of groundwater samples.

| Test No. | Test type | Purpose |
|---|---|---|
| (1) | T-test | To identify statistical difference in $NO_3^-$ concentration and EC between the three zones of the city (south, central, north). |
| (2) | Regression analysis | To correlate groundwater contamination with population density and sewerage coverage. |
| (3) | T-test | To indicate any deterioration in groundwater quality between '93 and '97. Deterioration in quality is reflected by a statistically significant increase in indicator ($NO_3^-$ and EC) levels. |

## 3.4 Results

### 3.4.1 Groundwater

The chemical analysis of groundwater from the sampled wells show that the concentration of pollution indicators differed between the three zones of the city, with the highest concentrations in the central zone (Figure 3.2, and Table 3.6). Table 3.7 shows the compliance of groundwater in the sampled wells with the WHO drinking water guidelines for $NO_3^-$ in each of the city's zones. The percentage of wells exceeding the WHO drinking water guidelines for $NO_3^-$ increased over time.

Table 3.6 Variation in groundwater quality in the city of Sana'a.

| Indicator parameters | South | Central | North |
|---|---|---|---|
| | **Sampling session 1993** | | |
| EC (µS/cm) | 460–590 | 380–2050 | 440–690 |
| $NO_3^-$ (mg/l) | 14-36 | 1–140 | 1–15 |
| | **Sampling session 1997** | | |
| EC (µS/cm) | 450-600 | 360-226 | 450–720 |
| $NO_3^-$ (mg/l) | 14-40 | 2-290 | 6–17 |

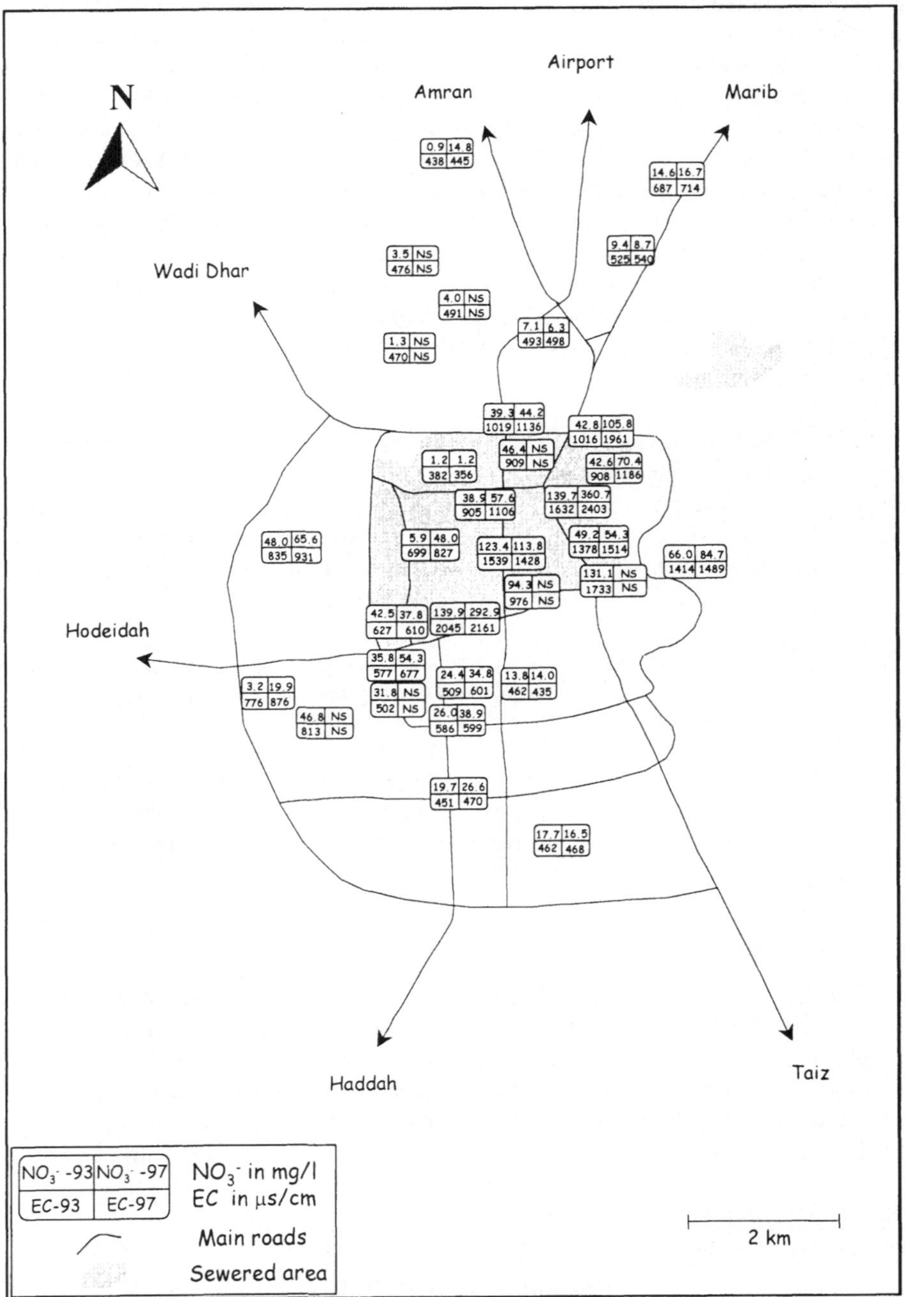

Figure 3.2 Nitrate and EC concentrations of groundwater from sampled deep wells.

Table 3.7 Groundwater quality compared to the WHO drinking water guidelines for $NO_3^-$.

| Zone | Sampling session | Total number of sampled wells | Number of wells with $NO_3^- > 45$ mg/l |
|---|---|---|---|
| South | | 9 | 1 |
| Central | | 16 | 9 |
| North | | 4 | 0 |
| **Total** | 11/1993 | 29 | 10 (34%) |
| South | | 7 | 1 |
| Central | | 13 | 9 |
| North | | 4 | 0 |
| **Total** | 7/1997 | 24 | 10 (42%) |

The results of the first statistical test suggest a statistically significant difference in nitrate concentrations between the three zones of the city for both '93 and '97 sampling sessions (Table 3.8). As for EC, the results show a significant difference between the central zone and both southern and northern zones, but no significant difference was found between the southern and northern zones in either of the sampling sessions.

Table 3.8 Statistical test (1): Results of a t-test reflecting the spatial state of groundwater contamination under the Sana'a city. $V_I$: set of observations, i: no.

| **Variable** | | **Observations** | | **Mean** | | **T-test** | | **Test outcome** |
|---|---|---|---|---|---|---|---|---|
| $V_1$ | $V_2$ | $V_1$ | $V_2$ | $V_1$ | $V_2$ | $T_{statistical}$ | $T_{critical}$ | |
| **$NO_3^-$ -93** | | | | | | | | |
| South | Central | 9 | 16 | 24.3 | 65.7 | -3.40 | 2.09 | Sign. Diff. |
| South | North | 9 | 7* | 24.3 | 5.8 | 3.97 | 2.20 | Sign. Diff. |
| Central | North | 16 | 7* | 65.7 | 5.8 | 5.19 | 2.12 | Sign. Diff. |
| **$NO_3^-$ -97** | | | | | | | | |
| South | Central | 7 | 13 | 29.3 | 102.8 | -2.50 | 2.16 | Sign. Diff. |
| South | North | 7 | 4 | 29.3 | 11.6 | 3.96 | 2.30 | Sign. Diff. |
| Central | North | 13 | 4 | 102.8 | 11.6 | 3.14 | 2.18 | Sign. Diff. |
| **EC-93** | | | | | | | | |
| South | Central | 9 | 16 | 571 | 1126 | -4.57 | 2.09 | Sign. Diff. |
| South | North | 9 | 7* | 571 | 511 | 1.08 | 2.16 | No Sign. Diff. |
| Central | North | 16 | 7* | 1126 | 511 | 5.26 | 2.10 | Sign. Diff. |
| **EC-97** | | | | | | | | |
| South | Central | 7 | 13 | 592 | 1316 | -4.13 | 2.13 | Sign. Diff. |
| South | North | 7 | 4 | 592 | 549 | 0.52 | 2.31 | No Sign. Diff. |
| Central | North | 13 | 4 | 1316 | 549 | 4.36 | 2.14 | Sign. Diff. |

* Concentration of three additional NWSA wells was included in the test as to increase the number of observations in the northern zone.

The results of the second statistical test (multiple linear regression analysis) are shown in Table 3.9. The analysis correlates contamination levels with variables that assumed to be the cause of the contamination i.e. population density in the well's immediate vicinity (as an approximation for wastewater discharge quantity), and presence of sewerage

Table 3.9 Statistical test (2): A stepwise multiple linear regression analysis to relate groundwater contamination to population density and sewerage coverage.

| Entered independent variables | Excluded independent variables | R | $R^2$ | Std. error of the estimate |
|---|---|---|---|---|
| **Dependent variable ($NO_3^-$ -93)** | | | | |
| Population density<br>Sewerage coverage | -<br>Sewerage coverage | 0.63 | 0.400 | 27.3 |
| **Dependant variable ($NO_3^-$ -97)** | | | | |
| Population density<br>Sewerage coverage | -<br>Sewerage coverage | 0.58 | 0.302 | 50.6 |
| **Dependent variable (EC-93)** | | | | |
| Population density | - | 0.572 | 0.327 | 332.6 |
| Sewerage coverage | - | 0.676 | 0.458 | 304.9 |
| **Dependent variable (EC-97)** | | | | |
| Population density<br>Sewerage coverage | -<br>Sewerage coverage | 0.584 | 0.310 | 419.7 |

The results of the third statistical test to test the hypothesis of an increase of indicator concentrations over time is summarized in Table 3.10.

Table 3.10 Statistical test (3): To test the significance of a temporal increase of indicator concentrations in groundwater under the city of Sana'a.

| Variables | Mean | Observations | Pearson correlation | T-test | | Test outcome |
|---|---|---|---|---|---|---|
| | | | | $T_{Statistical}$ | $T_{Critical}$ | |
| **$NO_3^-$ 93 – $NO_3^-$ 97** | | | | | | |
| $NO_3^-$ 93 | 39.7 | 24 | 0.899 | -2.46 | 2.07 | Significant increase |
| $NO_3^-$ 97 | 66.18 | 24 | | | | |
| **EC 93 – EC 97** | | | | | | |
| EC 93 | 849 | 24 | 0.922 | -2.62 | 2.07 | Significant increase |
| EC 97 | 977 | 24 | | | | |

### 3.4.2 The distribution network at Nokom

Despite the fact that some disturbances were encountered during sampling Session 1, its results are presented in Figure 3.3 for comparison and discussion purposes.

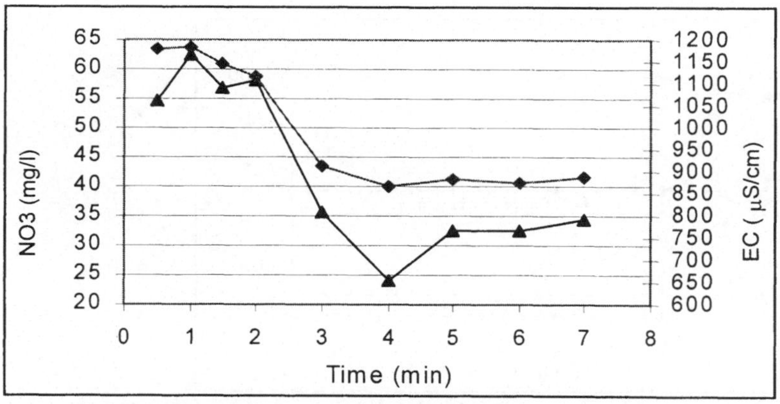


Figure 3.3 Nitrate (▲) and EC (◆) variation in the distribution network at Nokom (Session 1) after opening of supply valve (t=0).

In Session 4, the absence of a trend in the EC values (520-550 μS/cm, which falls within the background range of the supplying well) led us not to conduct nitrate analysis for this session. Nevertheless, suspected contaminated samples (slight visual discoloration) were analyzed for ammonium (Table 3.11).

Table 3.11 Ammonium concentration of selected samples of Session 4.

| Sample Number | $NH_4^+$ (mg/l) |
|---|---|
| 1 | 0.09 |
| 4 | 0.529 |
| 6 | 0.284 |
| 7 | 0.671 |
| 9 | 0.194 |
| 10 | 0.00 |

Analytical results of the undisturbed Sessions 3 and 5 are shown in Figures 3.4 and 3.5.

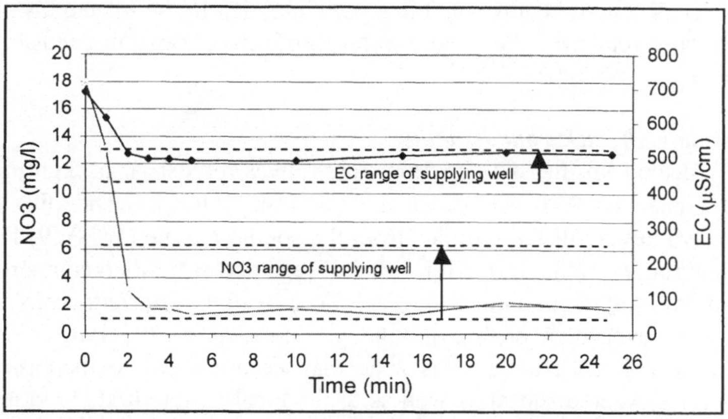


Figure 3.4 Nitrate (▲) and EC (◆) variation in the distribution network at Nokom after opening of supply valve (t=0) (Session 3).

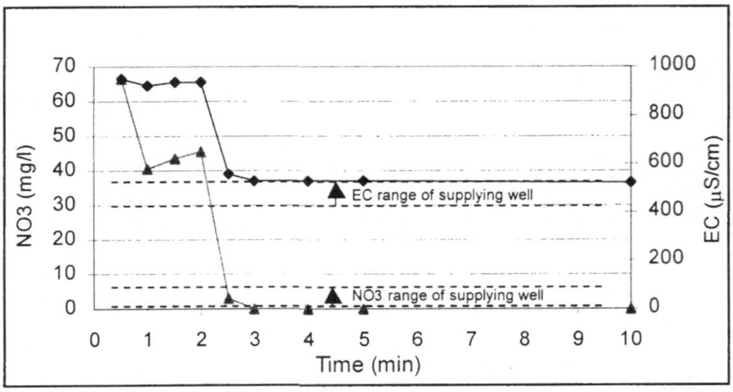


Figure 3.5 Nitrate (▲) and EC (◆) variation in the distribution network at Nokom after opening of supply valves(Session 5).

## 3.5 Discussion

### 3.5.1 Groundwater

In order to be able to conclude that the direct disposal of domestic wastewater into the sub-soil layers has resulted in groundwater contamination, the background water quality, and the representativeness of the sampling program need to be known. Assuming that the NWSA well inventory, which included around 140 deep wells, covered only 50% of the total actual number of private wells within the city, the sample size of 29 relatively randomly selected wells is around 10%, which may be considered representative. Although the different zones of the city were not covered equally in the number of sampled wells, they were selected to reflect the different population densities within the city. Even though the northern part of the city was under-represented, water quality data of three NWSA production wells located to the north of the city are used to compensate this deficiency. Given that the groundwater system in Sana'a consists of three aquifer layers in addition to a complex local geology, it is quite difficult to determine with certainty the aquifer from which a well taps. Although the results cannot be attributed to a particular depth, the selected wells are assumed to generally represent the deeper, more productive Tawilah sandstone aquifer.

#### *3.5.1.1 Comparison with background quality*

The NWSA wells located in the neighboring rural areas reveal a background groundwater quality range of 330-780 µS/cm and 0-14 mg/l for EC and nitrate, respectively. The EC values of all the samples from the southern zone (W1-W9) fell within the background range (Table 3.6). Although still regarded safe when compared with the WHO drinking water guidelines (Table 3.7), all nitrate concentrations of samples from this zone exceeded the background groundwater quality range. The relatively higher nitrate concentration in this zone may be attributed to increasing reliance on cesspits for sewage disposal as well as to the local agricultural activities. In the middle zone of the city (W10-W25), 70-80% of the sampled wells exceeded the background range for EC. As for nitrate, initially ('93 sampling session) three wells contained nitrate within the background range, but this number decreased to only one well in the '97 sampling session, while the rest of the wells in this zone exceeded always, and often the background quality range and the WHO drinking water quality of all samples from the northern zone (W26-W29) fell within the background range for both EC and nitrate. Water quality data of the three NWSA production wells (P16, P18 and P19) (Figure 3.1) located to the north of the city, with close proximity to the sampled wells, indicate also that their concentrations fell within the background range for both EC and nitrate.

Despite the fact that the water levels of the Tawilah aquifer are deep (100-130 m below ground surface), the prevailing contamination may be explained as follows:

- A suspected interaction between the shallow alluvium groundwater aquifer and the deeper Tawilah sandstone aquifer could induce downward movement of contaminated shallow groundwater.
- Cesspits in Sana'a are quite deep (10-25 m), thus reducing the thickness of the unsaturated zone.
- In many cases, proper grouting of private boreholes is neglected during construction, allowing short-circuiting (downward flow of contaminated water along the borehole wall), thus creating point sources of pollution.
- High hydraulic loading on cesspits (160-280 mm/d), which is the result of small pit diameter (average 1.5 m) and the high number of inhabitants per household (average 7).

It is suspected that the wastewater discharged to the underground is gradually diluted with groundwater, and partially pumped back up for use within the city. Although in need for further investigation, the slow groundwater flow in Sana'a suggests that the horizontal spread of pollution due to groundwater flow will take a time several orders of magnitude longer than vertical transport due to the high abstraction rate within the city.

#### *3.5.1.2 Spatial variation of groundwater contamination*

A statistically significant difference in nitrate concentrations exists between the three zones of the city, with the northern zone having the lowest concentrations followed by the southern and central zones, respectively. As for the EC, a similar pattern to that of nitrate existed, though no significant difference between the southern and northern zones was found. These results suggest higher contamination levels of groundwater in the central part of the city, which, is thought to be proportional to, among other factors, the intensity of domestic wastewater disposal.

When investigating the correlation between population density and sewerage coverage, and the groundwater contamination, it was found that the population density played a larger role in explaining groundwater contamination. Although, it might be expected that groundwater contamination correlates inversely with sewerage coverage, the situation in Sana'a differs inasmuch that the sewered area does not necessarily eliminate the use of cesspits as connection to the sewerage system was only obligatory for those households with cesspits that had to be destroyed as they were located along the sewer lines. This means that within the sewered area cesspits are still in use. Moreover, with conversion to sewers being relatively recent, the underground layers are still saturated with wastewater. Finally, the area of infiltration may be much wider than the selected zone, and percolation maybe not vertical.

The regression analysis indicates that the population density explains the local groundwater contamination with $R^2$ values of 0.30-0.40. Although this is low, it is considered reasonable given the relatively small sample size and the effects of other factors that influence the degree of contamination, such as the local geology.

#### *3.5.1.3 Temporal variation of groundwater contamination*

A statistically significant increase was observed in the concentrations of pollution indicators between 1993 and 1997. Although the time span between the two recordings is relatively small (4 years) given the slow nature of groundwater movement, higher concentrations may indicate a higher superficial rate of wastewater loading (discharge). In order to conclude with certainty that a temporal increase in groundwater contamination is taking place, a long-term monitoring program should be installed with appropriate sampling frequency.

Topography limits expansion of the city to the east and west, thus forcing rapid urbanization in the southern and northern directions. In combination with governmental financial constraints to expand the sewerage system, the growing reliance on cesspits will probably lead to an increase in groundwater contamination in those parts of the city as well.

A large portion of the city's population has lost its confidence in the quality of groundwater supplied by either the public or private water supplies. This is reflected in the rapid increase in drinking water purchases from the growing number of unregulated small (private) filtration stations, with average water prices of 10 and 100 times higher than that of the public and private bulk supplies, respectively (chapter 2). Consequently, public and private "groundwater originated" bulk supplies are mostly used for non-potable uses.

### 3.5.2 The distribution network at Nokom

During normal operation of water distribution networks, one of the measures often taken to eliminate the risk of contamination (back-siphonage or cross-connection) is to maintain continuous water supply with adequate water pressure. For intermittent supply systems, this safety feature does not apply, and the daily fluctuation in water pressure between negative and +60 mwc can damage the joints, fittings and connections in the network, thus further increasing the risk of contamination. This risk becomes even higher when the network is laid out in an area saturated with corrosive wastewater.

Although samples of Session 4 were not analyzed for nitrate (used throughout this investigation as the main contamination indicator), the presence of ammonium in the analyzed samples demonstrates contamination since the supplying well contained no ammonium. The ammonium content in the samples is still low if compared with the typical range of 90-200 mg/l in the raw sewage of Sana'a (chapter 7), which suggests a large dilution in the network and/or nitrification within the soil layer.

Although results of Session 1 demonstrated contamination of the network at the sampling point, the subsequent discovery of a large break in the supplying main raised the question whether similar contamination would occur during normal operation. While the contamination pattern of Session 1 cannot be used to confirm network contamination under "no-damage" conditions, it can be used for comparison of results obtained later in Sessions 3 and 5. In Session 1 (Figure 3.5) EC and $NO_3^-$ of the initial sample were 34% and 100% higher than the subsequent constant quality samples, collected 3-7 minutes later. Unlike all other sampling sessions, samples of this session tested positive for both total and fecal coliforms.

Results of Sessions 3 and 5 (Figures 3.4 and 3.5) show a pattern similar to that of Session 1, with both nitrate and EC concentrations dropping to within background levels after the first three minutes of water supply (opening of the main valves). This suggests that any contaminated water in that particular part of the distribution network is flushed out within the first three minutes of continuing water supply. Having proven that the pattern of the results in session 1 confirms contamination[41], and in order to establish that the pattern of sessions 3 and 5 indicate also comparable contamination, a cross-correlation analysis should reveal the similarities between the results of the three sessions (Table 3.12).

Table 3.12 Cross-correlation between the results of Sessions 1,3 and 5.

| | EC Session 1 | EC Session 3 | EC Session 5 | | $NO_3^-$ Session1 | $NO_3^-$ Session 3 | $NO_3^-$ Session 5 |
|---|---|---|---|---|---|---|---|
| $EC_{Session1}$ | 1.0 | - | - | $NO_3^-{}_{Session1}$ | 1.0 | - | - |
| $EC_{Session3}$ | 0.81 | 1.0 | - | $NO_3^-{}_{Session3}$ | 0.68 | 1.0 | - |
| $EC_{Session5}$ | 0.98 | 0.76 | 1.0 | $NO_3^-{}_{Session\ 5}$ | 0.88 | 0.85 | 1.0 |

Table 3.12 shows a strong correlation between the results of Session 1 and that of Sessions 3 and 5, thus confirming a similar contamination either through breaks in the network pipes or through joints and fittings.

The initial samples of Sessions 3 and 5 had relatively higher nitrate concentrations, which for Session 5 exceeded the range of 15-40 mg $NO_3^-$/l for the raw sewage in Sana'a (chapter 7). However, as the contamination is suspected to be diluted, this suggests that the $NO_3^-$ concentration of the contaminating sewage is exceptionally high, thus far exceeding that of raw sewage. In combination with the very low ammonium concentration, this suggests a high nitrification activity during the passage of the sewage through the soil, which is quite likely. Unless the exact dilution rate of the infiltrated (sucked in) sewage to the water supply pipe is known, the nitrogen and other balances cannot be further quantified, however.

The higher ratios of both EC and $NO_3^-$ concentrations in Session 5 in comparison to those of Session 3 suggest a higher rate of contamination (Table 3.13). This, in turn, may suggest deterioration of the network during the 1.5 years elapsed between Session 3 and Session 5.

Table 3.13 Ratio of initial samples' EC and $NO_3^-$ concentration to the average background value.

| **Session** | **EC** | **$NO_3^-$** |
|---|---|---|
| 3 | 1.3 | 6 |
| 5 | 1.8 | 22 |

[41] Due to the subsequent discovery of a break in the supplying pipe.

The results show a strong likelihood of widespread contamination of the distribution network in the Nokom area. It can also be argued that these results can be extrapolated to other city zones with similar circumstances.

## 3.6 Conclusions and recommendations

Widespread wastewater disposal through cesspits to the underground has deteriorated to a substantial degree the quality of groundwater under some parts of the city of Sana'a. This pattern of groundwater contamination correlates well with population density, as groundwater under the central zone of the city tends to be more contaminated than that under the less densely populated southern and northern zones. Much of the contaminated groundwater is being pumped up again to the city for further use, while the remaining part is expected to follow the general groundwater flow towards the north. Nevertheless, the slow horizontal movement of the deeper groundwater and heavy pumping suggest a slow horizontal spread of pollution.

The large fluctuations in water pressure in the distribution network as a result of intermittent supply in the area of Nokom systematically induces back-pressure and suction of wastewater from the sub-soil into the network through any fissure in the pipe or its fittings, thus contaminating the water in the network.

Areas facing failure of on-site wastewater disposal facilities should be sewered by priority to minimize the health hazards associated with contamination of water supply and direct exposure to overflowing wastewater from on-site sanitary facilities.

## 3.7 References

Al-Agha M. (1995) Environmental contamination of groundwater in the Gaza Strip. *Environmental Geology*. **25**, 109-113.

Al-Hamdi M. (1997) Case study XIII Sana'a, Yemen. In: Helmer R. and Hespanhol I. (eds.). *Water pollution control*, E&FN Spon, London.

Anderson Y. (1991) A waterborne disease outbreak. *Wat. Sci. Tech.* **24**(2), 13-15.

Anderson Y. and Stenstrom T. (1987) Waterborne outbreaks in Sweden-Causes and etiology. *Wat. Sci. Tech.* **19**, 575-580.

Bucksteeg K. (1991) Water resources management and water pollution control: National case studies from the Federal Republic of Germany. *European Water Pollution Control*. **1**(4), 34-44

Craun G. (1985) A summary of water born illness transmitted through contaminated groundwater. *Journal of Environmental Health*. **48**(3), 122-127.

Craun G. (1991) Causes of waterborne outbreaks in the United States. *Wat. Sci. Tech.* **24**(2), 17-20.

Farid M., Atta S., Rashid M., Munnink J. and Platenburg R. (1993) Impact of the reuse of domestic wastewater for irrigation on groundwater quality. *Wat. Sci. Tech.* **27**(9), 147-157.

Featchem R., McCarry M. and Mara D. (1977) *Water, Waste and Health in Hot Climates*. Wiley, London.

Flipse W., Katz B., Lindner J. and Markel R. (1984) Sources of nitrate in groundwater in a sewered housing development, central Long Island, New York. *Groundwater*. **22**(4), 418-426.

Foppen J. (1996) Evaluation of the effects of groundwater use on groundwater availability in the Sana'a basin, Main report. Technical Report No.5, vol. 1, SAWAS-3, National Water and Sanitation Authority, Sana'a, Yemen.

Gerba C. (1987) Transport and fate of viruses in soils: field studies. In: Rao V. and Melnick J. (eds.). *Human viruses in sediments, sludge and soils*. CRC Press, New York.

Gerba C. and Rose J. (1990) Viruses in source and drinking water. In: McFeters G. (ed.) *Drinking water microbiology progress and recent development*, Spring-Verlag, New York.

Gondwe E., Nwanuzi F. and Mbwette T. (1997) Septic tank-soakpit systems in Dar es Salaam, Tanzania. *J of Environmental Engineering*. January issue.

Ijzerman M., Hagedorn C. and Reneau R. (1992) Fecal indicator organisms below on-site wastewater system with low pressure distribution. *Water, Air and Soil Pollution* **63**, 201-210.

Lahti K. and Hiisvirta L. (1995) Causes of waterborne outbreaks in community water systems in Finland: 1980-1992. *Wat. Sci. Tech.* **31**(5-6), 33-36.

Lerner D. and Tellam J. (1992) The protection of urban groundwater from pollution. *Journal of the Institution of Water and Environmental Management*. **6**(1), 28-37.

Marzouk Y., Goyal S. and Gerba C. (1979) Prevalence of enteroviruses in groundwater in Israel. *Groundwater*. **17**, 487-491.

Nashashibi M. and Van Duijl L. (1995) Wastewater characteristics in Palestine. *Wat. Sci. Tech.* **32**(11), 65-75.

NWSA (1996) Evaluation of the effects of groundwater use on groundwater availability in the Sana'a basin. Technical Report No.5, vol. 5, SAWAS-3, National Water and Sanitation Authority, Sana'a, Yemen.

Paul J., Rose J. and Jiang S. (1997) Evidence for groundwater and surface marine water contamination by waste disposal wells in the Florida Keys. *Water Research*. **31**(6), 1448-1454.

Rao V. and Melnick J. (eds.). *Human viruses in sediments, sludge and soils*. CRC Press, New York.

Rashed M., Awad S., Salam M. and Smidt E. (1995) Monitoring of groundwater in Gabal El Asfar wastewater irrigation area (greater Cairo). *Wat. Sci. Tech.* **32**(11), 163-169.

Shariatpanahi M. and Anderson A. (1987) Bacterial survey of well water-Tehran, Iran. *Environmental Research*. **43**, 285-289.

Slade J. and Edworthy K. (1981) Virology of wastewater recharge of the chalk aquifer. II. Microbiology and water quality. In: Goddard M. and Butler M. (eds.). *Viruses and wastewater treatment*. Pergamon Press, New York.

*Standard Methods for the examination of water and wastewater* (1980). 15th edn, American public health association, New York, USA.

Wiesenekker H. (1996) Microbiological groundwater contamination of the Sana'a city area. Technical note No. 23, SAWAS-3, National Water and Sanitation Authority, Sana'a, Yemen.

William S., Schiff S. and Robertson W. (1984) Chemical fate and transport in a domestic septic system: unsaturated and saturated zone geochemistry. *Environmental Toxicology and Chemistry*. **13**(2), 193-203.

Zoller U. (1993) Groundwater contamination by detergents and polycyclic aromatic hydrocarbons-a global problem of organic contaminants: is the solution locally specified?. *Wat. Sci. Tech.* **27**(7-8), 187-194.

# Chapter 4

# Water Ownership and Conflict Resolution

## 4.1 Introduction

Legal systems of many countries nowadays include water resources in the public domain and thus treat them as public property. This may be justified by the importance of water to sustain life and its importance to the economy and hence the welfare of the population. Importantly, with water being a common-pool resource, an argument exists for stricter regulation of its ownership (Chapter 1). For a long time, ownership of groundwater in western countries was governed by the *Roman Rule* and the *English Rule* giving the land owner the absolute ownership of the underlying water (Solanes, 1999). In the Islamic system, ownership of groundwater is more of a communal nature, but may be subject to private ownership when labor or capital input is undertaken to harness the water. Thus, a person who constructs a well is entitled to the water that percolates to his well (Caponera, 1973; Wilkinson, 1978). These rules are challenged by the rapid advances in drilling and pumping technology leading in many cases to unsustainable use (aquifer overdraft) and externalities such as drawdown, salination, baseflow reduction, and land subsidence. The depletive effects of groundwater in some states of India has been argued to be essentially the result of the absence of a clearly defined property rights system that specifies upper limits of pumpage for both individual and collective abstractions (Saleth, 1996). The state of Texas, USA, still employs the *English rule*[42] of absolute ownership to groundwater. Todd (1992) argues that employment of such a law, which endorses ownership based on capture but at the same time fails to recognize liability for aquifer drawdown, creates limited incentive for efficient and environmentally safe abstraction. Although Texans are sensitive to any state infringement on their private property rights, the increasing rate of groundwater depletion of some aquifers has forced the state to move towards greater regulation (Wurbs, 1995).

In the absence of a regulatory framework that sets, monitors and enforces limits on groundwater abstractions, absolute private and communal ownership of groundwater do not provide security to the users. This is because everyone overlaying an aquifer has the right to abstract water and consequently nobody enjoys a long-term secure water source, the only secure water being pumped water, thus encouraging users in many cases to exceed the safe yield of the aquifer.

In the absence of a water law in Yemen, groundwater management has been largely left to the users (domestic (both public and private), industry, and agriculture). Although the Constitution declares all natural resources to be state property, the absence of a detailed legal framework, the lack of proper institutional setup to administer and manage water resources, and the weak capacity of the government to implement and enforce water related policies have given to the local users the

[42] Often refers to as the "law of the biggest pump".

possibility to unsustainably abuse many fossil aquifers. This situation is typical in the Sana'a basin, where heavy groundwater abstractions for domestic and mostly for agriculture has lead to an annual drop of 3-6 meters in water level in many regions of the basin area (Laes and Bamatraf, 1991). Unlike surface water irrigation (spate and spring), where traditional detailed customary rules have compensated for the nonexistent official management of water resources, large-scale groundwater irrigation was only made possible after the introduction of drilling and pumping technology in the 1970s, thus very few customs regarding current groundwater use actually exist. Groundwater management to control overuse is complicated by four main factors, of which the first two are of a general nature and the second two are more particular to the Sana'a case. The first is the wide availability of groundwater, but of which the condition is dependent on the action of many users. Thus, unless each individual can be assured that others are behaving in a manner to protect the resource base, the *in situ* value of the resource becomes low, promoting higher use rate. The second factor is the adoption of indirect policies, i.e., subsidies on energy, which may distort the incentive to use groundwater. The third factor is the nature of the crops cultivated within the Sana'a basin area, particularly qat, which is a highly demanded crop with high financial returns thus promoting expansion and eliminating any incentive to reduce water application to lower the cultivation costs. The fourth factor is the unique tribal social and power structure of the area, combined with the general mistrust that exists between the government and the tribes, which increases the difficulty level for official control over groundwater use.

The agronomic and financial incentives to use groundwater for irrigated agriculture are examined in Chapters 5 and 6. This Chapter examines the legal and cultural status of groundwater in the Sana'a basin. The objective is to clarify the official stand of the government as well as the perceptions of the local farmers towards the issues of groundwater ownership and use rights. Moreover, given the high dependence of the rural population of the basin area on groundwater for irrigation[43] and the high depletion rate of the aquifer, it is also important to clarify and understand the resolution mechanisms of water related conflicts. This understanding would be beneficial in identifying the key actors in the decision making process of groundwater use.

## 4.2 Methodology

The approach followed to carry out this part of the research depended mainly on collection of relevant data from various sources (books, reports and publications) and obtaining opinions, views and perceptions of farmers through a field survey. Being the largest water user within the Sana'a basin, the survey concentrated on irrigated agriculture, by interviewing 79 farmers from the different agricultural regions of the basin area (for more details on the size and selection procedures of the survey sample, refer to Chapter 5). As for the purpose of this Chapter, the questionnaire included questions pertaining to the relevant issues of groundwater ownership as well as contingent valuation-type questions related to the reaction of farmers towards possible governmental actions[44] to control aquifer overdraft. In addition, direct questions about

[43] Cultivation of cash crops is the main economic activity of the rural population of the basin area.

[44] These actions were: (1) introduction of pumping tax, or volumetric charge, (2) Physical limitation to well yield, and (3) an increase in energy (diesel) prices.

the different resolution mechanisms of water related conflicts were included in the questionnaire.

Greater insight into the issues of groundwater ownership, water rights and conflict resolution was also sought through formal and informal discussions that took place with many farmers, particularly those in the vicinity of the monitored "pilot" farm[45] area, during the two-year fieldwork period.

## 4.3 Results and discussions

### 4.3.1 Legal status of water ownership

#### *4.3.1.1 State view*

Governed by the sources of the legal system in Yemen (Islamic law (*shari'ah*), legislation and customs), water ownership is interpreted in different ways in the Yemeni legal sources. The *shari'ah,* which is incorporated in the Civil Law, considers water in its natural environment as a non-salable commodity to which every person within the community has a right. The Constitution, however, establishes water resources as the property of the state, which should oversee their utilization and exploitation in such a way that the public welfare is served (Article 8). As the Constitution regards water resources to be state property, and stipulates that a law must be issued to organize and manage their use, the current exploitation of the aquifers can be interpreted as an informal concession by the government to the present users.

Although a national water law has been drafted twice, it has not been deliberated in the parliament, reflecting the sensitive and complicated nature of water related issues particularly in a water short region like Yemen. As with regard to the Sana'a basin, the government realized early in the 1970s the fragile and depletive nature of the aquifer and reacted by passing a presidential decree (law No. 14, 1973) that established protection zones around the public domestic urban water supply well fields. The decree introduced a permit system as well as setting upper limits for groundwater abstractions for private uses within these protection zones. Although these actions may theoretically be viewed as effective deterrents for private expansion in groundwater use, four overlooked factors made this decree ineffective.

- The designated protection zones were relatively small in comparison to the total basin area.
- The tribal social structure of the area posed a challenge for the government to effectively enforce the decree, thus leading to many illegal drillings. The decree seems to have been only enforced on weak villages (tribes). As an interviewed farmer explained "previously we were denied permission to drill wells, but since "Naji"[46] became the tribal sheik, we now drill wells by force". This area falls within the protection zone.
- Despite the fact that the decree established upper limits to pumping, the lack of any measuring devices on private wells made such rule redundant and ineffective.

[45] Refer to Chapter 5 for more details on the pilot farm.
[46] Sheik Naji is also a member of Parliament.

- All investment to develop groundwater for irrigation had been drawn from private sources, so it did not seem logical or acceptable for farmers to be denied access to groundwater.

Decreeing water resources to be state property was the first step towards control over scarce groundwater by creating protection zones within the Sana'a basin. However, to make these measures operational and effective on the ground they should have been followed by broad political commitment[47] among the stakeholder groups. High transaction costs[48] and the low enforcement capacity of the central government hinder the effectiveness of the permit system, making it practically ineffective.

As for the role of the local administrative councils in water resources management, the recently passed Local Authority Law directs the local elected district councils to:

- Observe and monitor the implementation of the national water policies and protect water resources from depletion and pollution according to the relevant laws and policies.
- Develop water resources by promoting further construction of structures and protect these resources from depletion and pollution according to the findings of scientific studies.

With the absence of clear national water policies, it is difficult to comprehend that the local councils can perform these directives or duties, particularly with no national water law in place.

*4.3.1.2 Shari'ah and customary views*

Though all Islamic schools agree that water placed in a private receptacle or container becomes a private property, three out of the four main schools consider wells as receptacles. According to the *shari'ah* principles, water that has been harnessed by means of capital or labor inputs, i.e., digging a well, belongs legally to the person who undertook that enterprise (Wilkinson, 1978). As with regard to the salability of water in the Islamic system, the most restrictive view proclaims that water selling is absolutely prohibited, but the prevailing view is that water selling is only prohibited as long as it has not been appropriated. In other words, water can become a private property (and become salable or tradable) when it is carried or transported to a private tanker or receptacle (Caponera, 1992; Al-Eryani et al., 1996).

The heavy groundwater use in the Sana'a basin cannot be considered a customary practice since it was only introduced in the 1970s after the introduction of modern drilling and pumping technology. Thus, unlike for surface water[49], very few customs regarding groundwater ownership and use rights actually exist.

In the past (prior to the introduction of pumps) open dug wells were mostly used for domestic uses, but in some cases they were also used for supplementary irrigation of small gardens of fruit trees or qat. Although these wells were privately owned, the survey revealed that they were sometimes lent out to neighboring farmers for free. In

---

[47] Since passing the law, administrative procedures have been limited to issuance of permits and fee collection. In many cases, these procedures are not even enforced.

[48] Includes establishment of a basin monitoring unit to, inventory and register all wells, install meters, conduct routine meter reading, and collect fines.

[49] For further readings on customary water management of surface water refer to Maktari (1971), Varisco (1982), Vincent (1990) and Kohler (1996).

order for a farmer to borrow a well he had to obtain a written permission from the well owner as to safeguard and protect the property right of the original owner. This "no compensation" activity may be explained by the relatively low return from irrigation as compared to the high cost of obtaining the water (manual power, either human or animal). In other words, since the costs exceeded profits, well owners did not view the use of their wells as an economic activity and thus demanded no compensation. The situation nowadays is completely the opposite, where groundwater irrigation is viewed as a profitable economic activity leading to informal markets for buying and selling groundwater. Therefore, it may be concluded that the change in value of groundwater to the well owner has changed the customary practice of *well borrowing* to be replaced by the current economic activity of *water selling*.

Within the set of nationally shared customs, ownership of groundwater is only implicitly recognized in roles that mainly aim at minimizing externalities, i.e., the interaction between wells, by defining a protection radius around each well as equal to its depth. However, with advancements in drilling and pumping technology, nowadays a protection radius of 500 meters is the customary norm[50].

The *shari'ah* and the Civil Law allow the initiation of a diversion from non-appropriated water to be claimed for appropriation, even if it is taken from within the property of others (public or private)[51]. This is comparable to the "first in use - first in right" principle, but limited in quantity by sufficiency to the appropriator. Diversion of water from a source should not, however, harm the existing users, unless the water is taken for drinking or ritual ablution[51]. Although these conditions do not differentiate between surface and groundwater sources, for the latter the diversion right may be initiated by drilling a well[52].

Furthermore, the Civil Law grants the land owner the absolute ownership of what is above and beneath his land to any beneficial height and depth. It is also permitted to separate the ownership of the land surface from what is above or beneath it[53] as long as it does not contradict or conflict existing laws. The right to irrigation water is inheritable and allowed to be denoted in wills and in regions where it is not against the local customs, irrigation rights may even be rented out, but cannot be sold separately from the land[54]. For well and spring owners, losing the right to groundwater seems only applicable if the well or spring is fully depleted.

According to the *shari'ah,* water is the gift of God, thus wasteful use is forbidden; priority of use is given first to human drinking followed consecutively by livestock drinking and irrigation (Al-Sayaghi, 1982). It should however be noted that this priority system seems to be governed by the physical presence of the user at the water source. In other words, the well owner cannot deny a person drinking water, but may prevent the user from transporting water to his living area even if it is to be used for drinking. Therefore, it is not clear whether the right to drinking water, based on the priority system, holds equally true to the larger basin or macro level. Unlike for surface water, groundwater use for irrigation seems not to be confined to the source

[50] During the field survey, it was observed that this rule is widely violated.
[51] Civil law, article 1367.
[52] Civil law, article 1366.
[53] Civil law, article 1163.
[54] Civil law, article 1370.

area, and it is quite acceptable, as far as local customs are concerned, to pump groundwater from the wadi and use it elsewhere, probably within the boundary of the supply region (or tribal area) (Al-Eryani et al., 1996).

While the *shari'ah* principles coincide with the farmers' current short-term interests, the strict applicability of these principles may be questioned under situations of high water scarcity in the light of the technological advances permitting heavy groundwater mining. The *shari'ah* has two complementary concepts, which may be separated by the level of water scarcity. In the no-scarcity situations (around the well), water collected in wells (in this case comparable to a receptacle) is subject to private ownership, while in scarcity situations (basin wide), the priority-in-use concept can emerge as a guideline to determine the priority in the "re"-allocation of groundwater resources. Despite the apparent confusion on the issue of groundwater ownership, the *shari'ah* principles can provide guidance in situations of scarcity at basin scale, through the priority-in-use-rule. Vincent (1990) reported the Islamic law to be remarkably flexible and pragmatic in matters related to water management, succeeding in many cases to develop consensus among the different parties rather than creating public disputes. Wilkinson (1978) indicated also that the traditional and Islamic practices have generally been in harmony, in that the Islamic water law has been able to adapt to the local needs and circumstances.

#### *4.3.1.3 Users' views*

The survey results confirm the contradiction between the Constitution implying that water resources are state property, the *shari'ah* giving the land owner the freedom to drill a well and thus practically own any water inside the well, and the actual perceptions of users. The *shari'ah* principles seem to support the perception of farmers. This is likely not because of the extensive prior knowledge of farmers about such principles, but rather because they offer the farmers a rationale to extract rents from resource use. The inability of the central government to control groundwater use through strict enforcement of the permit system only reinforces the perception of the farmers that they are in their full right when depleting the aquifer.

Given the ineffectiveness of the official policy to control and manage groundwater in the Sana'a basin, the current on-farm practices and the perceptions of farmers gain higher importance since they are unchallenged and simply prevail. During the survey 80% of the respondent farmers responded that, according to them, ownership of groundwater is directly linked to land ownership. Twenty five percent of those, further restricted ownership by the type and place of use: groundwater should only be used for irrigation[55] of agricultural land within the boundary of the tribal region. The remaining 75% supported unconditional use of groundwater giving the land owner the freedom to do whatever he wishes with the abstracted water from his well (Figure 4.1).

It seems that there is some discrepancy between the perception of farmers when asked about ownership rules and what is actually taking place in practice. While a large number of the respondents confirmed that a direct link exists between water and land, in practice, actual ownership of groundwater is only proportional to the share one has in a given well, and thus the amount of withdrawn water is not necessarily governed

[55] Although small in quantity, use of groundwater for domestic uses within the region is also permitted.

by the size of the overlaying owned land. For example, a farmer with a sizable farm growing a low value crop would probably decide to have a smaller share in the supplying well than another farmer with a smaller farm growing a higher value crop. Land ownership may therefore be considered a criterion to permit access to groundwater without any limitation on the quantity to be abstracted, regardless of the size of the "attached" irrigated land.

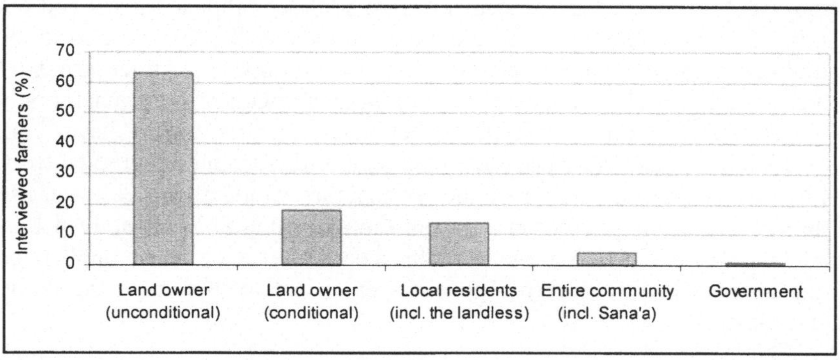


Figure 4.1 Farmers' perception of groundwater ownership (n=79).

Unlike in the Sana'a basin, in Arizona, USA, the quantity of groundwater that can be abstracted is made explicit and mostly based on *grandfathered rights,* the historic water use, which is commonly proportional to the size of the irrigated land (Kohlhoff, 1988). Intersectorial transfer of groundwater rights has been taking place in Arizona since the 1980s[56], but governed by sustainability concerns to lower groundwater abstractions to safe levels. Therefore one can observe similarities between the Sana'a and Arizona situations in the connection between water rights and land ownership and differences in the defined quantity to be abstracted, where in Arizona the quantity is better defined and is proportional to the land size or based on the prior abstractions. The situation in Sana'a, however, seems to be based only on the financial capacity of the farmers to drill or share access to groundwater.

Only 14% of the respondents indicated that groundwater belongs to all the residents within the local community, including the landless (Figure 4.1). Although the landless in this case may in principle be viewed to own part of the groundwater, to abstract it, they would have to acquire a share in a well, which represents a high financial burden that they cannot afford. Furthermore, only 4% of the respondents thought that the entire community, including urban residents of Sana'a, share the right to groundwater. Surprisingly, only 1% of the respondents thought that groundwater was the property of the state. Although in some of the surveyed regions it is required by law to obtain a permit to drill a borehole, permits are perceived by many farmers to be an income source to the issuing agency, but not as a concession from the government to the permit holder to access groundwater.

[56] According to Kohlhoff (1988), faced with water supply shortages, the cities of Phoenix, Tucson, Scottsdale, and Mesa in Arizona, USA, have been involved in transfer of groundwater from irrigated agriculture to urban use through purchasing and retiring 5700, 9113, 3400 and 4500 hectares of farm land, respectively.

To reveal the farmers' position towards government control over water resources, the survey included contingent valuation-type questions. Nearly all the respondents expressed strong rejection towards the idea of introducing a groundwater abstraction tax or volumetric charges. Moreover, most farmers (around 90%) rejected the idea of physical limitation to the quantity of abstracted groundwater. However, only one farmer rejected indirect measures such as an increase in energy prices. The explanation given for the high rejection of abstraction tax and physical limitation on abstraction was that, while farmers can refuse to pay any proposed taxes and could resist the installation of flow measuring devices on their wells, they cannot do much against indirect measures. Indirect measures, however do not only target farmers, but affect other groups as well and should thus solicit resistance from the general public. Moreover, indirect measures such as higher energy prices would have limited reduction effect on groundwater use due to the high profits from cash crop cultivation (Chapter 6). Only 14% of the farmers indicated that they would alter their water use in case of an increase in the price of energy, but the remaining 86% of the respondents confirmed that they would continue with the current level of water application[57]. Clearly the generated profits are too powerful an incentive to change the farmers' water use decisions.

### 4.3.2 Views on groundwater exhaustion

Farmers stated that their perception towards groundwater has changed significantly since the start of the drilling rush in the 1970s and 1980s. While more than 90% responded that in the past they viewed groundwater to be non-exhaustible and even considered it an "underground sea", only around a third of the respondents still have the same belief. This is an evidence of the growing awareness of the limited nature of the aquifer. This change in perception has come about mainly through observation of the continuous decline of water levels. It may be argued, however, that this new perception may exert a negative influence on the water abstraction by creating a competitive atmosphere for higher water use as every farmer tries to get the most of the aquifer before it totally depletes. The farmers' views towards the causes of the drop in groundwater levels can be grouped into three main categories (Figure 4.2).

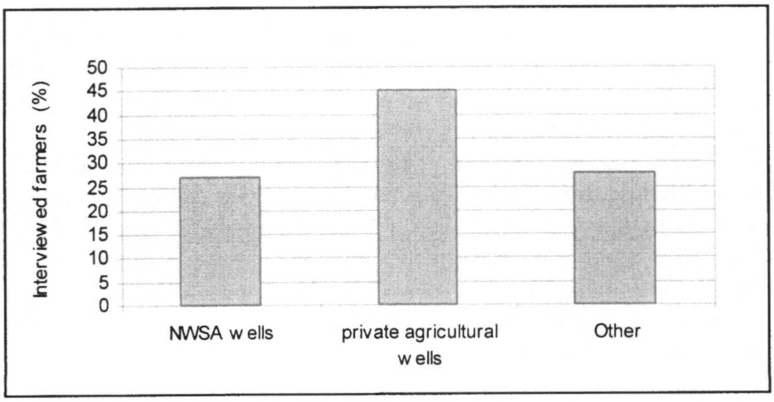


Figure 4.2 Farmers' perception on the cause of the drop in groundwater level.

[57] As one farmer put it, "an additional 2,000 YR for diesel is not much considering that a 60,000 YR harvest is on the line".

In the regions where NWSA has production wells nearly all respondents (27% of the sample) blamed NWSA wells for the decline in water levels. When the author objected that the number of the NWSA wells is small compared to that of the local farmers, the respondents justified their view by stressing the large production capacity of the NWSA wells. Twenty eight percent of the farmers blamed other reasons such as low rainfall and the absence of recharge structures (dams). The remaining 45% of the respondents admitted that groundwater pumping by the large number of private agricultural wells is the main cause of the drop in water levels; nevertheless, they continued to blame the government for not taking proper actions to construct more recharge dams.

### 4.3.3 Water availability and sharecropping

Sharecropping in the Sana'a basin is governed by two main criteria, namely the water source and the type of crops to be cultivated. These two criteria results in four main types of arrangements:

- Rainfed practices to cultivate cereals.
- Rainfed practices to cultivate cash crops.
- Groundwater irrigation to cultivate cereals.
- Groundwater irrigation to cultivate cash crops.

Before discussing these arrangements, it should be noted that the limited erratic rainfall in the area has created a special land use pattern, where every agricultural plot is attached to a designated rainwater harvesting land, which can not be sold separately and in some cases may be as much as 10 times the size of the plot itself. Therefore the value of an agricultural plot reflects usually the size of the attached rainwater harvesting land and thus the potential availability of irrigation water.

#### *4.3.3.1 Sharecropping of cereals*

For cultivation of cereals under rainfed practices, the land owner usually receives 1/3 of the harvest. In cases where cereal cultivation relies on groundwater for irrigation from the land owner's well, then the land owner would receive 1/2 of the harvest if he contributes 50% of the operational costs, or receives only 1/4 of the harvest in case he decides not to contribute to the operational costs of the well. In either case it is usual practice for the land owner to contribute 50% of the maintenance costs of the well. At the end of the contract period, the farmer does not become entitled to any of the cultivated land.

#### *4.3.3.2 Sharecropping of qat and grapes*

Unlike cereals, qat and grape cultivation requires greater attention especially at the plantation stage, where the first harvest is only possible after two or three years of nursing. Therefore, sharecropping of qat and grapes depends on whether the farmer is contracted to maintain an established farm or is asked to reclaim a new uncultivated land. In the case where the farmer comes into an established farm with a reliable water source, e.g., a well, then the land owner receives 1/4-1/3 of the harvest, while the farmer bears all operational costs and 50% of the maintenance costs of the well. In case reclamation of additional land takes place, the farmer becomes entitled to own 1/2 of the reclaimed portion of the land. In case the land owner contracts the farmer to initiate the establishment of a farm, the farmer is requested to bear all the necessary investment including an access to a water source (e.g., drill, or share in a well). After the second harvest, the farmer becomes entitled to part of the reclaimed land as compensation. The percentage of the land to be transferred to the farmer depends on whether the land owner is endogenous to the area, and on the type of the reclaimed land (agricultural plot or rainwater harvesting land). In case the land owner is not from the local village or area, the farmer becomes entitled to 1/2 of the reclaimed agricultural land and as much as 5/6 of the reclaimed rainwater harvesting land. In return, the original land owner becomes entitled to 50% of the farmer's share in the water source. In case the land owner is from the area[58], the farmer would only be entitled to 1/2 of the reclaimed agricultural land and only 1/3 of the reclaimed rainwater harvesting land. In return, the original land owner gains 50% of the farmer's share in the supplying well.

### 4.3.4 Conflict resolution of water related issues

In Yemen, there are basically two systems for conflict[59] resolution: the judiciary system and the arbitration system. The judiciary system is based on the law number 1, 1990, assigning courts with the responsibility for ruling in every litigation or crime. The arbitration system, on the other hand, is based on tribal customs, is mostly used in the rural areas and consists of several levels starting at the village/community level and ending at the tribal level (Al-Eryani, 1996). Within the tribal socio-political institution, mediation in water disputes by the village sheiks have been reported to be highly successful especially in the spring-irrigated regions of Yemen. By comparing the seasonal flood system to the spring system of irrigation, Varisco (1983)[60] concluded that the nature of the flow in an irrigation system determines the potential of water related conflicts within that region. The complexity of the allocation process when a large number of irrigators must act simultaneously in a limited period of time in the seasonal flood system creates a higher conflict potential than the spring system where little attention is needed for the supervision of the day-to-day activities. It can thus be suggested that the intensity of water related conflicts is inversely proportional to the reliability of and knowledge on the water flow. Therefore, the high reliability of groundwater for irrigation in the Sana'a basin may be comparable to that of the irrigation system using spring flows, thus explaining the observed low level of conflicts.

During the field survey, farmers were asked to explain the different resolution mechanisms relevant to water conflicts in the Sana'a basin area. Despite being in close proximity to the capital city (the largest urban area of the country), it was evident that the unofficial arbitration system is more prominent in water and land related disputes in the different regions of the Sana'a basin (Figure 4.3). However, many respondents stated that they had not been involved in such conflicts, but rather mentioned the path they would follow in case of a conflict, or cited paths that had actually been followed by acquaintances.

Conflicts can be resolved at any of the three identified levels (Figure 4.3). Only a very small percentage of conflicts would bypass arbitration and go directly to the official judicial system (path B). Around 19% of the respondents insisted that conflicts are

[58] Residency is not a criteria to establish endogenous status, but rather affiliation with and allegiance to the local tribe.

[59] Like in Vincent (1990), conflicts within this context are viewed to take place when the parties involved can not resolve the dispute among themselves and require the intervention of a higher authority or negotiator 'supra-group'.

[60] The fieldwork carried out by Varisco was conducted in the area of Al-Ahjur, 45 km east of Sana'a.

always resolved by arbitration at the tribal sheik level (path C+D), and even if not resolved the conflict would not advance to the official judicial system. Ninety six percent of conflicts are supposed to pass through the arbitration system (path A+C+D+E+F), and a large percentage of those (path E, F) would advance to the official judicial system in case the conflict persists. The reasons given for the popularity of the arbitration system were (1) the high costs associated with the judicial channels due mainly to corruption, (2) the long time it takes for a conflict to be resolved in courts, and (3) the general mistrust of the rural (tribal) population towards the government[61]. Moreover, judicial decisions are usually weakly enforced, due partially to corruption and the weak central government. Although the arbitration procedures are sometimes very costly as well, resolutions are normally accepted and enforced through social pressure given the strong influence of the tribal customs in the rural social structure.

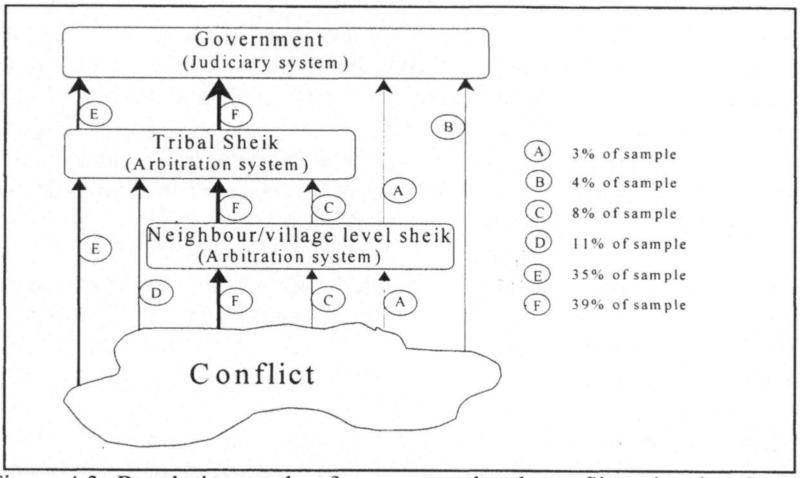


Figure 4.3 Resolution paths for water related conflicts in the Sana'a basin as perceived by farmers (n=79).

Given the dominant role of the *shari'ah* in legislation in Yemen, the religious elite was expected to have a great influence on water related matters. Surprisingly only 14% of the surveyed farmers considered the religious elite to have the necessary knowledge and competence to resolve water related conflicts. This is despite the heavy involvement of the religious elite in land allotment according to the Islamic inheritance system. The remaining 86% of the respondents indicated that farming in general and the subject of irrigation in particular is not the forte of the religious elite. On the other hand, 81% of the farmers identified their local parliament representative (who in some cases is the same village or tribal sheik) to have the necessary knowledge and ability to resolve water related conflicts.

Although the "sheik" was identified as an influential figure in the life of the rural population in the Sana'a basin, it is in the interest of the sheik to show concern in seeking and protecting the benefits and interests of the local population. For sensitive issues connected to the livelihood of the population, e.g., groundwater use for irrigation, sheiks are expected to defend rather than contest the actions of the farmers

[61] For further reading on the relation between the state and the tribe in Yemen, refer to Manea (1996).

(e.g., heavy groundwater abstractions), since he mostly draws his status, prestige and popularity directly from the local population. In this regard, the role of a sheik, as far as the local farmers are concerned, would not only be limited to stand against any governmental decision to control groundwater abstraction, but is also expected to lobby for projects aimed to enhance water availability (construction of dams). Like in the political arena, a divergent shift by the sheik from this expected role would probably result in lower popularity, which provides an opportunity for ambitious and wealthy persons to gain grounds and recognition as to become also sheiks. It is also essential to point out that these sheiks represent a principal interest group since, for the most part, they are farmers themselves, hence, have an additional incentive to contest and fight any governmental actions to control groundwater use.

## 4.4 Conclusions and recommendations

In Yemen, there is a discrepancy in the interpretation of the ownership of groundwater between the *shari'ah* and the Constitution. While the Constitution declares all natural resources to be state property, the *shari'ah* considers groundwater to be a communal property, but grants the well owner private ownership (tradable and salable) of all the water that percolates to his well. Given that heavy groundwater use is not a traditional activity, there are no detailed customary rules that govern or control groundwater use. According to the *shari'ah*, priority in groundwater use is given to human drinking followed consecutively by livestock drinking and irrigation.

Groundwater ownership and "re"-allocation should be further investigated through an in-depth examination of the *shari'ah* principles, particularly in light of the technological advances in pumping, which have lead to unsustainable use of groundwater in most regions of Yemen, but especially in the Sana'a basin.

Given the low capacity of the government to manage and control the use of groundwater, the views and perceptions of the local users (farmers) are essential. Most farmers in the Sana'a basin regard groundwater to be the property of the owner of the overlaying land, although access to groundwater is governed by the share in a well regardless of the farm size. Beside the weakly enforced permit system, there is no incentive to deter a Sana'a basin farmer from drilling a well in his land other than financial constraints, which is often overcome through sharing by neighbors in the capital and operational costs.

Perceptions of many farmers on the confined nature of the aquifer in the Sana'a basin has changed over the past two decades, as more farmers are becoming aware of the possibility of total depletion. This change in perception may have a negative influence on the attitude of the farmers, as every farmer tries to get the most of the aquifer before it totally depletes, thus creating higher level of competition.

The survey results display the tribe as a coherent socio-political institution with consistent framework for water allocation and for conflict resolution, and identify the sheik (both at the village and tribe level) as a key figure in decision-making and conflict resolution.

## 4.5 References

Al-Eryani M., Bamatraf A., Al-Sakkaf G. and Hadash S. (1995) Socio-economic and Water Rights Aspects. Technical note No. 17, SAWAS-3, National Water and Sanitation Authority, Sana'a, Yemen.

Al-Sayaghi H. (1982) *The Regulations of Sana'a in the Twelveth Century (AH)*. Dar El-Hekmah Press, Sana'a.

Caponera D. (1973) *Water Law in Moslem Countries*. Irrigation and Drainage Paper 20/1, FAO, Rome, Italy.

Caponera D. (1992) *Principles of Water Law and Administration: National and International*. Balkema, Rotterdam.

Ismail A. (1998) Socio-Economic consistency report. Ministry of Agriculture and Irrigation, Sana'a, Yemen.

Kohler S. (1996) the demands and limitations of sustainable water use in arid regions: A discussion of "sustainable development" taking the Wadi Markhah (Republic of Yemen) as an example. *Applied Geography and Development*. **47**, 25-36.

Kohlhoff K. (1988) Urban use of Arizona's rural groundwater. *Journal American Water Works Association JAWWA5*. **80**(3), 46-50.

Laes D. and Bamatraf A. (1991) Review of the Sana'a Groundwater Monitoring Network. Technical Secretariat of the high Water Council, Sana'a, Yemen.

Maktari A. (1971) *Water rights and irrigation practices in Lahj. A study of the application of customary and shari'ah law in south-west Arabia*. Cambridge University Press.

Manea (1996) La tribu et l'Etat au Yémen. In: Mondher Kilani (ed.). *Islam et changement social*. 205-218.

Saleth R. (1996) *Water Institutions in India: Economics, Law, and Policy*. Commonwealth Publishers.

Solanes M. (1999) Institutional and legal issues relevant to the implementation of water markets. In: Salman S. (ed.). *Groundwater: legal and policy perspectives*. World Bank technical paper no. 456. The World Bank. Washington D.C.

Todd D. (1992) Common resources, private rights and liabilities: A case study on Texas groundwater law. *Natural resources Journal*. **32**, 233-263

Varisco, D. (1982) Sayl and Ghayl: The ecology of water allocation in Yemen. *Human Ecology*. **11**(4), 365-383.

Vincent L. (1990) The politics of water scarcity: Irrigation and water supply in the mountains of the Yemen Republic. ODI/IIMI Irrigation management Network Paper 90/3e.

Wilkinson J. (1978) Islamic water law with special reference to oasis settlements. *Journal of Arid Environments*. **1**, 87-96.

Wurbs R. (1995) Water rights in Texas. *Journal of Water Resources Planning and Management*. **121**(6), 447-454.

# Chapter 5

# Analysis of Water Use in Irrigated Agriculture in the Sana'a Basin

## 5.1 Introduction

The low rainfall (200-250 mm/a) and the lack of surface water sources in the Sana'a basin led in the past to almost total dependence on dry farming practices for agriculture. The reliance on the erratic rainfall for irrigation configured both land use and cropping patterns. The limited availability of water led to a special land use arrangement in which every agricultural plot was connected to an attached catchment area along hilly slopes for rainfall harvesting. These catchment areas were never cultivated and were legally linked to the agricultural land, thus could not be sold separately. Agricultural practices were mostly of a labor-intensive nature to enhance the moisture holding capacity of the soil. The traditional cropping pattern was dominated by cereal crops like wheat, sorghum, and barley. Cash crops (qat, grapes, fruit trees and vegetables) were rarely cultivated under dry farming practices since they demand greater water quantities and for some crops, such as grapes, high initial financial investment.

The introduction of drilled boreholes in the early 1970s facilitated an easy access to a reliable water source that could be used for irrigation. As a result, many farmers invested in deep boreholes and thus converted from dry farming practices into irrigated agriculture. Sharing the investment for drilling is very popular among farmers within the Sana'a basin. It helps to distribute costs and serves as a framework to overcome land fragmentation. In many regions of India, Pakistan and Bangladesh the existence of informal water markets is argued to have promoted and enhanced equity and productivity of groundwater, allowing the poor to access a reliable irrigation water source (Shah, 1993). In the Sana'a basin area, the tribal social structure creates a strong social bond among the indigenous residents of a certain area serving the same purpose of facilitating collective action. This is reflected in the high level of cooperation among farmers resulting in a "cooperate well" sharing arrangement, where ownership of boreholes is not restricted to large and rich farmers, but rather includes the middle income and poor farmers as well.

In order to maximize the returns on their investment, farmers changed the cropping pattern in the 1970s-1980s from the low-profit cereals into high-profit cash crops. The main reasons that discouraged farmers to continue with a cropping pattern dominated by cereals were (i) the government policy to subsidize imported grains, thus lowering the market price of cereals, (ii) the rapid growth of Sana'a, increasing the demand for cash crops, particularly qat, and (iii) high profits from cash crops and low pumping costs resulting partially from subsidized energy. Within the basin area it is estimated that, at present, groundwater irrigated cereal cultivation covers less than 5% of the agricultural land in the basin (own observation).

The expansion in irrigated agriculture resulted in heavy groundwater mining throughout the basin area, leading to an annual 3-6 meters drop in groundwater levels. The Tawilah aquifer, the main productive aquifer in the basin, is also the principal water source for the growing municipal water demand, particularly within the capital Sana'a (Chapter 2). It has been estimated that the water deficit[62] in the basin area was around 135 $Mm^3/a$ in 1990 and it is projected to reach 400 $Mm^3/a$ by the year 2010 (HWC, 1992a). Given the slow and limited recharge of groundwater, the irrigated agriculture and thus the livelihood of most of the rural population within the basin area, as well as the municipal water sector are at risk, especially given the high cost of the alternative water sources (Chapter 2). It can be argued that water in the agricultural sector is not being utilized in an economic productive manner, because of the high opportunity value for its use in other sectors. To consider partial reallocation of groundwater from irrigated agriculture to municipal use requires the analysis of three sets of key issues that describe the demand structure for water in the agricultural sector:

- The agronomic characteristics of the irrigated agricultural sector in the basin area (land use, cropping pattern, water use, etc.).
- The financial and economic characteristics of the irrigated agricultural sector, as the incentive framework for decision making on water use.
- The perceptions and views of farmers towards water and the risks associated with its availability as factors that also affect decision making on water use.

This chapter will mainly deal with the first set of issues, and specifically analyze groundwater use for qat and grape cultivation in the Sana'a basin. The second and third sets of issues are dealt with in Chapters 6 and 4, respectively. The current Chapter will examine (1) accessibility of farmers to groundwater, including the physical characteristics that constitute accessibility to groundwater, i.e. well depth, water level, draw down and well yield as well as the drilling regulations and the ownership structure of boreholes; (2) water use by farmers in reference to actual farm-level water application and the identification of key factors that influence water application; and (3) the effects of water availability on land use in agriculture within the basin area.

## 5.2 Approach and methodology

### 5.2.1 General

One important source of data were the reports of the water and agriculture projects conducted in the Sana'a basin. As water use for irrigated agriculture is a relatively recent development and since cash crop cultivation, particularly qat, is controversial, comprehensive factual information from those reports was very limited and questionable due to the lack of field data. Therefore, during this research, field data were collected through two approaches. First, a representative typical (cash crops) pilot farm was selected to be monitored in detail for three complete irrigation seasons ('96,'97, and '98). Second, a broader but less detailed field survey of a representative sample of cash crop farms from within the Sana'a basin area was conducted during the '97 irrigation season. The survey data reflected also still the '96 situation.

[62] This is based on the estimated 42 $Mm^3$ of natural recharge.

### 5.2.2 Pilot farm survey

The pilot farm was selected based on the following criteria:

- Representativeness: Consistent with the following criteria, the pilot farm should be as much as possible a typical farm, representative for the practices in the basin.
- Location: The farm should be located in the agricultural area preferably in the proximity of the city.
- Cropping pattern: Since irrigated agriculture is dominated by cash crops, the farm should have a mix of cash crops, preferably qat and grapes representing the most important crops requiring most water.
- Size: The farm should be average in size (1.4-2.3 hectare) (Central Bureau of Statistics, 1993). Moreover, the number of plots on the farm should not be too large for accurate data recording.
- Water source: From several preliminary visits to the area and interviews of various farmers, it was concluded that most farmers depend on shared wells as the source for irrigation water. Therefore, the pilot farm should rely on a shared borehole.
- Owner cooperation: It was quite important for the owner of the selected farm to be willing to be monitored and thus provide realistic data. This should include the installation of a water meter on the well. To obtain agreement, it required also the consensus of all the other shareholders.
- Assistance: Since monitoring the farm activities required continuous record taking, availability of a capable assistant (preferably a person who resided in the farm area) was essential for proper data collection.

After several visits to different regions of the basin area, a farm east of Al-Rowda, 10 km northeast of Sana'a along the Marib road, was selected (Figure 5.1). The farm consists of twelve plots of which six are planted with qat (Q1-Q6), and the remaining plots with grapes (G1-G6) (Figure 5.2). The total area of the farm is 2.14 ha, of which 0.34 ha is under qat and 1.8 ha under grapes. However, the effective area under grapes (excluding unplanted land between grape rows) is 0.78 ha. The farm is exclusively irrigated with groundwater from a borehole (drilled on the farmer's land) owned by five shareholders (Table 5.1).

Table 5.1 Shareholders and their shares in the borehole.

| Share holder | Share |
|---|---|
| Mr. Yaeash | 3/8 (has the right to three days in an eight-day cycle) |
| Mr. Al-Zendani* | 2/8 (has the right to two days in an eight-day cycle) |
| Mr. Al-Heilah** | 1/8 (has the right to a day in an eight-day cycle) |
| Mr. Hamzah | 1/8 (has the right to a day in an eight-day cycle) |
| Mr. Haj | 1/8 (has the right to a day in an eight-day cycle) |

* Owner of the selected pilot farm to be monitored.

** This share is further divided into three equal shares, resulting in a 1/24 share each.

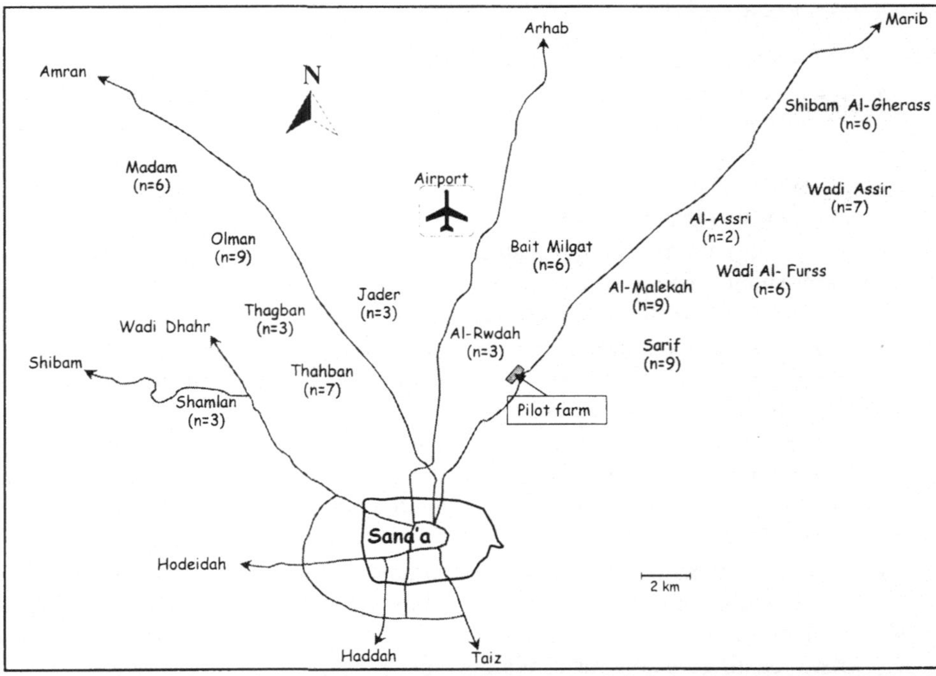


Figure 5.1 Location of the pilot farm and the surveyed farms within the Sana'a basin area (n= number of surveyed farms within each sub-region).

The supplying well was drilled in 1979 to a depth of 200 meters below ground level (mbgl). Prior to 1997, the well was equipped with a one cylinder 24-hp diesel engine, which generated a total yield of around 10 $m^3/h$. The static water level in the well was 138.6 mbgl in October 1995. It was not possible to measure the water level more regularly due to the lack of a separate observation tube.

In order to monitor water use, a water meter was installed in the delivery pipe to the selected farm. The delivery pipe is divided into two main branches, one towards the qat plots and the other towards the grape plots. Each branch is provided with several outlets and valves to control the direction of the flow and the exact discharge point. Plastic hoses are usually used as connectors between the various outlets and the different plots.

For practical reasons, all shareholders agree at the beginning of each irrigation season on a water distribution schedule to be closely followed during the entire season. After periods of shutdown of the well for maintenance, the same schedule resumes starting with the next shareholder in line. The first two shareholders of Table 5.1 are usually scheduled to utilize their shares in consecutive days. Although a schedule is pre-determined at the start of the season, it is quite flexible if two farmers want to switch their turns as long as others are not affected. The shares also determine proportional contribution to investment and operation and maintenance (O&M) costs and, thus, ownership.

After careful screening (mainly through direct observation) of the farmer's sons, one of them was selected as a full-time assistant. The selection was based on his education level (secondary school graduate, i.e. 12 years of basic education) and his thorough knowledge of the different aspects of farming. Moreover, the selected person was unemployed (spent most of his time on the farm), and most importantly, was in charge for the operation of the well.

Prior to the '96 irrigation season, the data collection forms were prepared and explained to the assistant. In total, four different forms were prepared to include (1) water application, (2) labor activities, (3) expenditures (purchase of diesel, oil, fertilizer, pesticides, clay, and maintenance), and (4) revenues (from crop sales). At the beginning of the '96 irrigation season, daily visits to the farm were undertaken to ensure smooth recordings, reduced later to 2-4 visits per week. These irregular visits provided also a chance to observe the various farming activities and an opportunity to crosscheck and control the quality and accuracy of the recorded data, as well as a chance to hold informal interviews, not only with the selected farm owner, but also with the other well shareholders and neighboring farmers. Although these discussions were informal, they provided valuable insight on the perceptions, views and experiences of farmers towards farming in general and water use in particular.

The collected information can be grouped into two categories. The first is directly related to water use and thus provides records on the quantity of water applied to each plot. The recorded data included water quantity as read directly from the water meter, as well as the application time, which served as a crosscheck of the meter readings. This category also included records of all labor work conducted within each plot. The second category covered all financial aspects of the farm, particularly all expenditures and revenues. Although members of the farmer's family[63] carried out most of the work within the farm, equivalent labor costs were determined using the average market compensation rate.

As a result of the declining yield of the well, the shareholders decided to invest in larger capacity equipment at the end of the '96 irrigation season. They purchased a new submersible pump, new suction pipes and a second-hand 6-cylinder diesel engine, thus increasing the yield by almost threefold to reach 30 $m^3/h$. This provided an opportunity to investigate water use under two different scenarios of water availability (or scarcity) ('96 vs. '97). At the end of 1997, the government raised the diesel prices by 67%. This rise in energy prices provided the opportunity to examine its effects towards water use by extending the monitoring program for a third season ('97 vs. '98).

[63] Mr. Al-Zendani has seven sons (10-30 years of age) who contribute to farming.

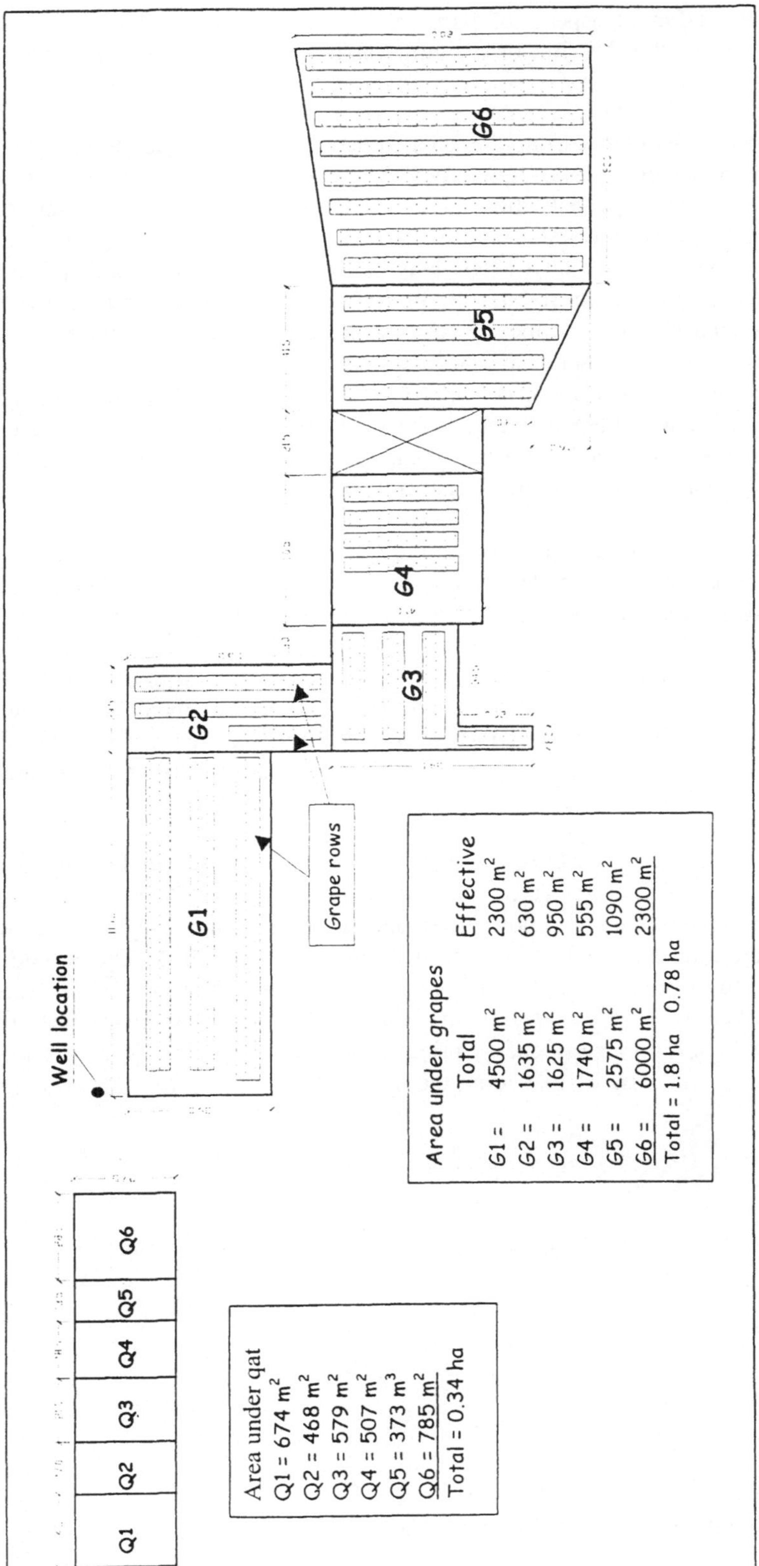


Figure 5.2 Layout of the pilot farm (dimensions in meter).

### 5.2.3 Field survey

#### *5.2.3.1 The questionnaire*

The questionnaire was designed to obtain three categories of information: (i) technical (water source (borehole), the farm, and water use), (ii) financial (expenditures and revenues of the farm), and (iii) farmers' perceptions. Farmers' perceptions, views and experiences, were acquired through questions on the different social, legal and political aspects of water and land use.

The questionnaire was prepared in English, then translated into Arabic by two different persons. Differences between the two translations were sorted out to yield simple direct questions. However, it was observed that the questionnaire was too long. The questionnaire was tested in the field for a period of one week during which it was modified by clarifying ambiguous questions and eliminating others. Even though the original questionnaire included detail questions about each plot within the surveyed farm, it became clear during the testing period that no reliable data could be determined this way and that it was time consuming. It was decided to modify the questionnaire as to concentrate on only one plot of each crop (considered representative by the farm owner) within the selected farm. Overall, the questionnaire was shortened so that an average interview would take around 45-60 minutes, after which farmers were observed to lose interest. The survey was conducted in the period of July-August 1997 and was designed to collect data of the '96 irrigation season. In special cases where farmers reported abnormal circumstances during the '96 season, data of what they considered a normal year were also recorded. Realizing that farmers do not keep exact records of their activities and given the nature of the solicited data, many replies were of "best memory" and "best estimate" nature. Great care was taken to check and improve the consistency of the replies within each farm survey as well as among the farm surveys (a copy of the questionnaire is included in Appendix 1).

#### *5.2.3.2 Sample selection*

Since wells are better inventoried than farms, sample development was based on well, not farm selection. From the well inventory carried out by the Sources for Sana'a Water Supply (SAWAS) project, the wells were grouped in categories according to their yield, from which a 5% (80 wells) stratified random sample was selected. Although the yield of all selected wells had been measured during the inventory, information like owner name, well depth, water level, etc., were missing for many wells. Once the wells to be surveyed were selected, a work plan was prepared to conduct the field visits, as the objective was to conduct interviews with the farm owners supplied by these wells. The field visit schedule included at least four interviews daily. Methodological problems that arose during field testing included:

- In many cases only the SAWAS well number with location coordinates were available, but in the field the location could only be estimated since no instrument was available to pinpoint the exact well location. Moreover, the only available map (1:50,000) was not detailed enough.
- The SAWAS number on many wells had faded out, or as was found out later, had disappeared since they were printed on walls that had been demolished or on a truck's door that had been sold.
- When a pre-selected well was identified, in many cases none of the shareholders would be nearby. Looking for them in the neighborhood was found to be time consuming. Among those successfully tracked down, some

felt suspicious and uncooperative since they were being specifically asked for by name.

Therefore, it was concluded that conducting the survey according to the plan would be time consuming and might result in too inaccurate information. It was decided to alter the well and farm selection procedures as to conduct around six interviews in each of the 13 major agricultural regions of the northern part[64] of the basin area. Sample selection within each region was conducted by visiting meeting points (fields, watch towers, qat sessions, etc.), explaining the purpose of the visit and asking for respondents. Although this selection method may not totally fulfill the requirements for random selection, it was found to be an effective and appropriate way to obtain consistent and coherent data. Visits were usually carried out at around 2-3 p.m. and interviews would last throughout the afternoon. Eventually 67 wells and 79 farms were surveyed of which 49 farms contained both qat and grapes, while 25 and 5 farms contained only qat or grapes, respectively (Figure 5.1). The 67 visited wells not only supplied the 79 surveyed farms with water, but also supplied additional unsurveyed farms. It was not the objective of the survey to inventory all the farms supplied by each well because by doing so, geographical representation of the entire basin would not be possible since a handful number of shared wells would cover the intended survey target of 70-80 farms.

Since the modified sample selection was no longer based on the SAWAS well inventory, the yield of the wells supplying the selected farms had to be measured manually. To measure a well yield, a marked 50-liter container and a stopwatch were used. Measurements were repeated 2-3 times to obtain an average reading. In some instances where a well was not in operation (commonly for maintenance purposes), the yield would be measured during a second visit.

It was observed that some farmers were reluctant and afraid to give out information at first. However, once the objectives of the survey was fully explained (i.e., academic research, and that the surveyor was from the university and not connected to the government), farmers would be more cooperative. It was also explained that cooperation was voluntarily, and would not lead to any further consequence. This clarification was necessary since the research subject (water and qat) is of a highly sensitive nature. Nevertheless, in few cases where the respondents were too disinclined and possibly unreliable, the interviews were canceled. Although often a local person accompanied the surveyor, almost all of the interviews were carried out by the surveyor for the following reasons:

- Consistency in the recording and interpretation of the obtained information.
- Time constraints (visits were conducted after normal working hours between 2-7 p.m.)
- Financial constraints.
- Lack of reliable and trained assistants.

With three interviews conducted daily, most regions were visited at least twice, and the entire survey took around two months.

---

[64] The northern three districts of the basin area, Hamdan, Bani Husheish and Bani Al-Harith are suspected to consume > 90% of the total agricultural water demand within the entire basin area (HWC, 1992a)

The software package *SPSS* was used to determine the correlation between the different sets of data, i.e., theoretical irrigation water requirements and the actual on-farm water application as well as between the monthly irrigation water application of the monitored farm during the different seasons. These correlations were conducted for both qat and grapes to test the hypothesis that while grapes are under-irrigated and given higher priority in irrigation scheduling, qat is over-irrigated but allows greater flexibility in irrigation scheduling.

## 5.3 Results and discussions

### 5.3.1 Accessibility of groundwater for irrigation

#### *5.3.1.1 Drilling activity and ownership of boreholes*

Figure 5.3 shows the age distribution of the 67 surveyed wells, reflecting the well drilling activity between 1970-1997. The trend follows a three stage development pattern, which may be characterized as the "initial stage" (1970-1980), the "rush stage" (1981-1990) and the "declining stage" (1991-1997). During the initial stage financial resources were quite limited and farmers were reluctant to invest in the new and untested technology of borehole drilling. Once the technology proved viable and successful, the second stage followed with farmers trying to obtain financial resources. This stage coincided with large monetary transfers from migrating Yemeni labors in the Gulf States. Some farmers acquired the necessary funds to invest in boreholes by selling part of their agricultural land.

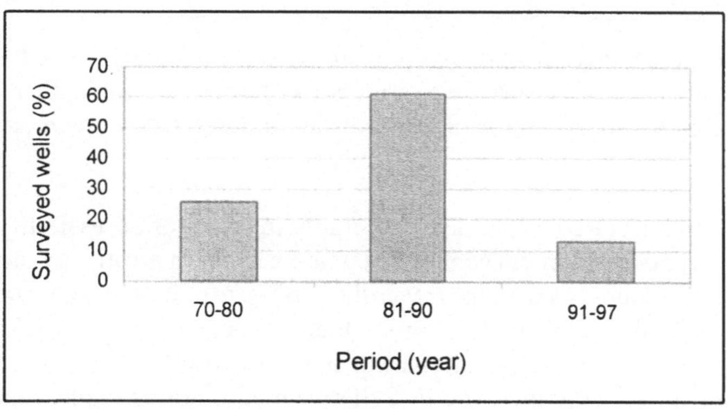


Figure 5.3 Age distribution of the surveyed wells (n=67).

Despite the high financial return from irrigated agriculture, the third stage shows a slack in the drilling activity due to two main reasons: (1) land constraints, as most suitable land had been developed, and (2) financial constraints, due to the devaluation of the local currency (higher capital servicing cost) and drying up of international remittances.

Irrigated agriculture in the Sana'a basin is characterized by high dependence on shared wells (Figure 5.4). While each of the 43% of the surveyed farmers own only a share of 14% or less of their supplying well, 16% of the farmers own a share larger than 88% of their supplying well, which in practical terms means that these are exclusive owners. Around 38% of the farmers own between 14% and 50% of their supplying well. This important feature of water use within the basin area can be justified by the following main factors:

- Land fragmentation, which is a common feature of the traditional inheritance system.
- Scarcity of suitable land for qat cultivation. Land reclamation for qat cultivation is expensive and is usually of a small scale that does not justify the high cost of drilling a new private well, thus providing an incentive for sharing.
- Although location requirements for grape cultivation are not as strict as for qat, the high initial capital investment coupled with the relatively lower return does not justify the exclusive ownership of wells, thus providing a similar incentive for sharing.

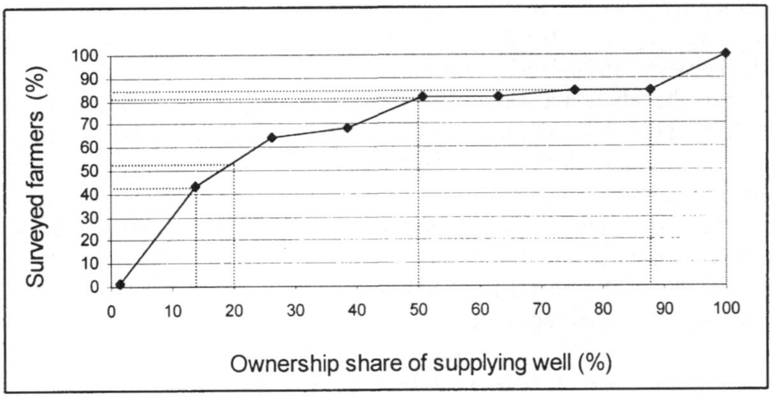


Figure 5.4 Cumulative percentage of ownership share in wells in the Sana'a basin (n=76) (each of the 43% surveyed farmers (34 farmers) owns a share of 14% or less in the supplying well; 16% of the farmers own a share larger than 88%).

Ownership sharing has also ensured a high level of equity in terms of accessibility to a reliable water source, as poorer farmers cannot be denied a share in a well as long as they fulfil the criteria of land ownership (regardless of size), indigenous status (tribally connected) and the ability to cover their financial responsibilities proportional to their share. The widespread sharing in wells results also in a high portion of the local population being involved in well ownership. For example, Figure 5.4 shows that 52% of the surveyed wells have 5 or more shareholders (households), representing a minimum of 35 inhabitants[65]. The high involvement of the local population in well ownership, however, could create an obstacle for the government to regulate groundwater abstractions as all decisions need to be consented with by all shareholders. This is further complicated by the fact that farming is the main economic activity within the area.

In four major regions of India[66], 74% of the net area under lift irrigation was reported to have been served by singly owned lifts while only 20% and less than 1% was served by jointly and co-operatively owned lifts, respectively (Saleth, 1996). The survey in the Sana'a basin revealed that while 36% of the irrigated area was served by singly owned wells, 64% of the land is served by the "cooperate-well" scheme, with an average ownership share of only 22% per shareholder. While land fragmentation in India, Pakistan and Bangladesh, is evident from the intensive behavior of the water market, where many sellers are also buyers themselves, in Sana'a, land fragmentation is apparent from the intensive state of well sharing, where shares as low as 1% were reported during the survey. In addition to land fragmentation, the comparatively high financial effort to drill in hard rock is likely to contribute to a high level of well sharing as well. Saleth (1996) reports a relatively higher percentage of land irrigated by jointly owned lifts in the hard-rock regions, with a range of 17-40% compared to 0-2%, and 3-5% in the mountainous and the Deltaic regions, respectively. This indicates that in regions with unfavorable hydro-geological conditions, like in Gujarat, India, and in Sana'a, the higher cost of drilling serves as a second incentive for well sharing. The much higher level of well sharing in the Sana'a basin may further be explained by the very deep groundwater levels (higher cost to drill and more expensive pumping equipment) as well as by the local social structure. Shah (1997) reports that in areas of India where the cost to obtain groundwater is high such as in the state of Gujarat, water extraction mechanisms are mostly owned by "tube well companies"[67], commonly selling water to members and non-members alike. While Saleth (1996) reports that 49% of the total pumpsets in four regions of India[68] were used by single farms, in Sana'a the survey revealed that only 16% of the wells were used by single farms. It thus may be concluded that the growing water scarcity reflected in high abstraction cost provided a decisive incentive for higher cooperation levels between farmers in the Sana'a basin.

The emerging informal groundwater markets reported in India, Pakistan and Bangladesh served to promote higher equity, which enhances income distribution benefiting mostly the poor and middle income farmers (Fujita and Hossain, 1995; Shah, 1997; Easter et al., 1999). Similarly, the high level of well sharing in Sana'a may be argued to serve the same purpose. Second, the more reliable access to irrigation water increased the productivity of irrigated agriculture, and in water scarce regions may have resulted in better efficiency through shifts in cropping patterns to more valuable crops increasing the value of groundwater. Nevertheless in areas facing groundwater overdraft, both intensive well sharing as well as groundwater markets fail to provide an incentive for resource conservation, especially in the absence of separate property rights on groundwater and the lack of proper policies to deal with such situations. Therefore, in areas with limited non-renewable groundwater resources, higher equitable access to groundwater may enhance equity, productivity and even efficiency, but usually at the expense of resource sustainability.

In the Sana'a basin, the very large number of stakeholders in water use on one hand complicats any effort to regulate abstraction. On the other hand, as this network of shareholding coincides to some extent with tribal structures, this may open avenues to improve awareness and facilitate collective action to reduce water abstraction.

#### *5.3.1.2 Physical characteristics of agricultural wells*

With an average total depth of 230 mbgl, the surveyed wells in the basin had a range of 100-350 mbgl. During the survey no open dug wells were observed to be in use for irrigation. Fifty percent of the surveyed wells had a depth less than 220 mbgl (Figure 5.5). While 72% of the surveyed wells range in depth between 150 and 300 meters, 19% and the remaining 9% of the surveyed boreholes had a depth greater than 300 m and less than 150 m, respectively.

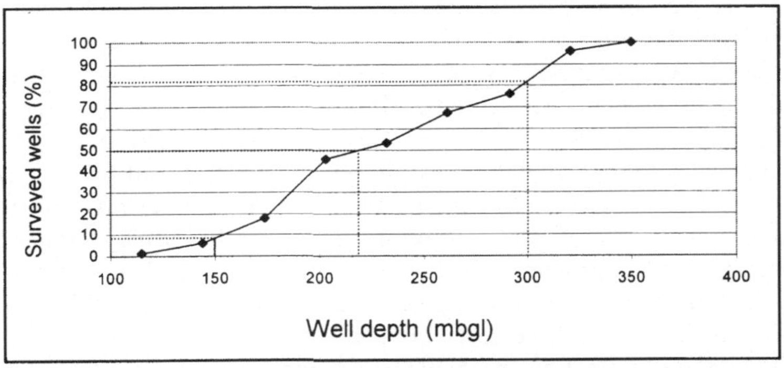


Figure 5.5 Cumulative percentage of the (total) depth of surveyed boreholes in the Sana'a basin (n=67 wells).

As for the drop in water levels, the average drop in groundwater level was 3.9 m/a with a range of 0-14 m/a. Only 9% of the surveyed wells had not experienced any drop in groundwater level, yet 77% had a drop in the range of 1.7-6.9 m/a (Figure 5.6). The remaining 14% is equally divided between the higher range (>6.9 m/a) and the lower range (< 1.7 m/a).

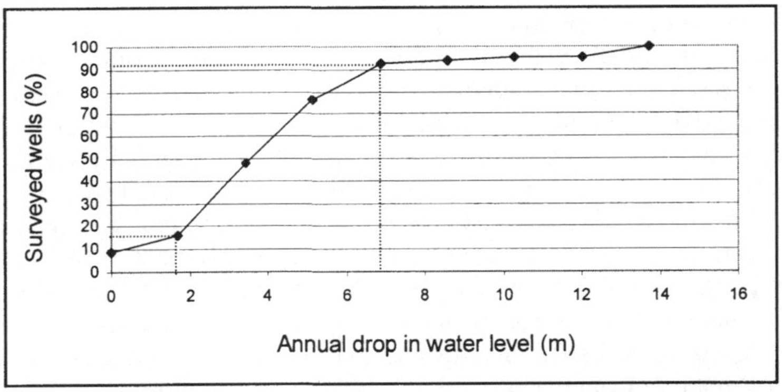


Figure 5.6 Cumulative percentage of the annual drop in groundwater level of the surveyed boreholes in the Sana'a basin (n=67).

Depth to the groundwater table of the surveyed boreholes ranged between 90 and 215 mbgl (Figure 5.7). This large variation is partly due to the topography of the different regions within the basin area. Another reason for this variation is the complex geological setting of the area in combination with heavy groundwater mining taking place in localized parts of the basin. Distinct areal variation in the depth to the water level occurred, for example the depth to groundwater level in the area of Shibam Al-Gherass ranged 165-210 mbgl with an average of 187 mbgl compared to a range of 93-126 mbgl and an average of 110 mbgl in the area of Al-Malekah. Both areas are located along the Marib road and are only 8-10 km apart.

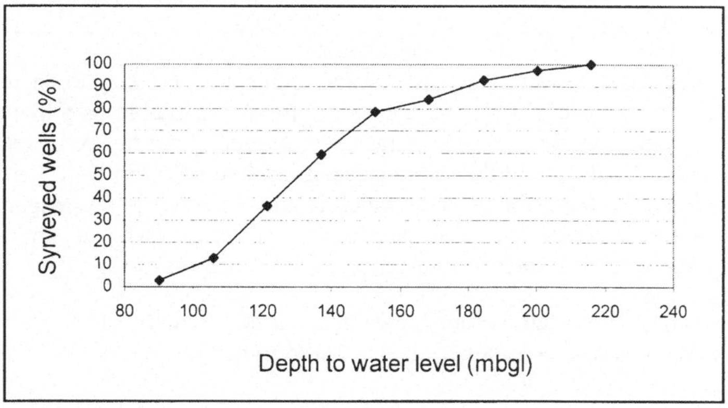


Figure 5.7 Cumulative percentage of the depth to groundwater levels of the surveyed boreholes in the Sana'a basin (n=67).

Although the yield of a well is very much determined by technical (well design) and hydrogeological factors, most of the wells within the basin were drilled by few contractors using the same drilling technology which explains the similar design and construction of many boreholes. The average yield of the surveyed boreholes was 4.4 l/s per well with a range of 1-12 l/s. Seventy three percent of the surveyed wells varied in yield between 2.3-7.5 l/s (Figure 5.8).

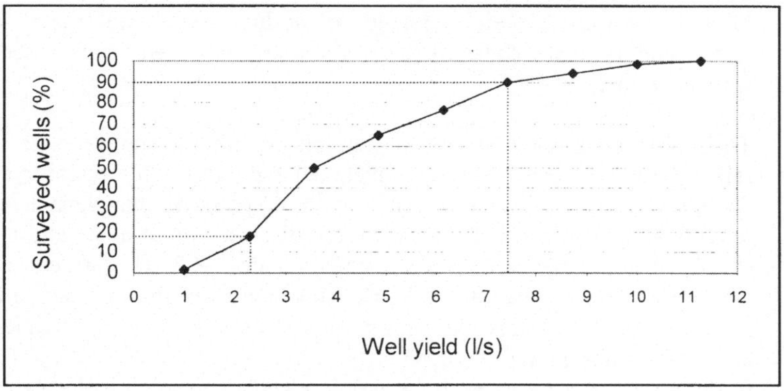


Figure 5.8 Cumulative percentage of well yield of surveyed boreholes in the Sana'a basin (n=67).

## 5.3.2 Groundwater use for irrigation

### *5.3.2.1 General*

In the Sana'a basin, the decision to irrigate is the responsibility of farmers who, in many cases, are illiterate and do not believe in scientific techniques to determine water requirements. Farm level decisions on irrigation scheduling and application are most often based on traditional practices and imitation of other farmers. Therefore, a large variation between the theoretical irrigation water requirements and actual water application was expected. The survey responses consistently revealed the same key factors that influence the farmer's decision making on water application for irrigation (Table 5.2).

Table 5.2 Factors influencing farmers' decisions on water application for irrigation in the Sana'a Basin.

| Factor | Outcome |
|---|---|
| Availability/ Accessibility of water | As accessibility to water decreases, water application is likely to decrease as well. Nevertheless, when dealing with a highly profitable crop like qat, farmers were found to purchase larger than needed quantities of irrigation water as accessibility to their water decreased (reduction in well yield or well shutdown for maintenance), which results usually in over-irrigation. This holds true both for well-owners and for farmers who do not own wells, but purchase water from neighboring well owners. |
| Risk aversion to ensure revenues | As revenues from agriculture increase, a farmer is inclined to bear the additional costs of extra water application to minimize the risk of a lower yield. |
| Agronomic information | It is widely believed among farmers that over-irrigation of qat can damage the quality of the harvest. This suggests also that other agronomic criteria (not further studied here) may influence the farmer's decision on irrigation water application. Farmers admitted that over-irrigation is sometimes used to compensate for insufficient manual work like soil plowing. |
| Customary and cultural beliefs | Though spillage and wastage of water is strongly discouraged in the local farming communities, over-irrigation is not considered wastage of water. Imitation amongst farmers with regard to habits in water application for irrigation has been observed regularly. Farmers tend to mimic successful farmers. However, their rationale to define "successful" and their understanding of the cause of this success, did not seem to be based on sound arguments. |
| Environmental concerns | Though water is scarce in Sana'a, farmers consider groundwater to be a reliable water source and many farmers believe that it will not completely deplete. The survey lends credence to the hypothesis that Sana'a basin farmers adopt escapist behavior towards "disaster" (total depletion of the aquifer) comparable to that described by Shah (1993) for farmers using groundwater for irrigation in India. This behavior find its supporting rationale in the following arguments, which were also frequently encountered during this survey:<br>(1) The farmer behaves as if he is completely unaware of the pending disaster and when it happens, he believes that he had nothing to do with it.<br>(2) He feels quite comfortable since many other farmers are equally affected. In case abstraction is regulated, he would continue to violate it while arguing that his overdraft alone (assuming everybody else complies) would not have a major damaging effect, yet if he complies while all other farmers do not, his water savings alone would not result in the salvation of the aquifer.<br>(3) In fact, he knows that depletion is imminent but feels that he should increase his pumpage before others do, leading to a more competitive environment, which results in a faster rate of depletion. |

*5.3.2.2 Irrigation water requirement*

Within the Sana'a basin qat and grapes are roughly estimated to consume around 40% and 25% of the total irrigation water consumption, respectively (HWC, 1992a). The irrigation water requirement is calculated by subtracting the effective rainfall from the crop water requirement. Crop water requirement is defined as the depth of water needed to meet the water loss through evapotranspiration (ETcrop) of a disease-free crop, growing in large fields under non-restricting soil conditions including soil water and fertility and achieving full production potential under the given growing environment. Therefore, ETcrop depends on (1) the local climatic conditions (temperature, humidity, wind, sunshine and evaporation), (2) crop characteristics (type of crop, stage of growth, growing season and the prevailing weather conditions), and (3) the local conditions and agricultural practices (distance and altitude, size of the fields, advection, soil water availability, salinity and irrigation and cultivation methods). Depending on the availability of relevant data, ETcrop for a particular crop is often calculated using one of four approximation methods (Doorenbos and Pruitt, 1984).

Using the Penman method, the irrigation water requirement for some crops within the Sana'a basin have been calculated (Table 5.3). Recent calculations suggest the irrigation water requirement for qat to range between 700 and 1,380 mm/a (Ting-Tiang, 1998). The approximate range of seasonal crop water requirement for grapes has been reported to be 650-1,000 mm/a (Doorenbos and Pruitt, 1984). Assuming an overall irrigation efficiency of 60%, the actual crop water requirement for grapes increases to 1,083-1,667 mm/a, and subtracting the effective rainfall, the total irrigation water requirement is expected to be somewhat lower.

Table 5.3 Irrigation water requirement for crops in the Sana'a Basin (After HWC, 1992a,b).

| Crop | Irrigation water requirement | |
|---|---|---|
| | **($m^3$/ha. season)** | **(mm/season)** |
| Sorghum | 7410 | 741 |
| Wheat | 6220 | 622 |
| Barley | 6220 | 622 |
| Maize | 7530 | 753 |
| Tomatoes | 10000 | 1000 |
| Potatoes | 9420 | 942 |
| Alfalfa | 16780 | 1678 |
| Grapes | 14000 | 1400 |
| Qat | 12050 | 1205 |

*5.3.2.3 Climatic and seasonal considerations*

The irrigation season in the Sana'a basin for qat and grapes depends mainly on the temperature and starts usually in February. Although for different reasons, both qat and grapes are viewed by the local farmers to be sensitive to the cold weather. Farmers believe that grape trees are affected to some extent by the severity of the previous winter; the colder it was, the higher the yield in the subsequent season. Nevertheless, a sudden drop in temperature after the start of the irrigation season is considered disastrous. Qat, on the other hand is not sensitive to cold temperatures as such, but to the combination of low temperatures during dew conditions. This combination, known locally as *tharieb*, can occur suddenly and results in desiccation of the trees, which consequently have to be felled. A major disaster of this nature occurred in the districts of Bani-Hushaish and Bani-Alharith in the Sana'a basin in the mid 1980s. It is for this reason that qat generally requires strict location criteria, and it is mostly found on protected plots (foot of hills and mountains) and not in large open fields.

It is widely known among farmers that qat is a durable crop that can withstand thirst. In principle, there is no distinct irrigation season and farmers can irrigate whenever they want to prepare the trees for pruning. Nevertheless, for the larger part of the Sana'a basin, qat is usually irrigated from February/March to October. Unlike qat, farmers consider grapes to have a distinct irrigation season that has to be closely followed to achieve an acceptable yield. The irrigation season for grapes starts usually in February, and depending on variety may extend into August or September. Results from the pilot farm as well as from the survey show that grape plots are given priority over qat in irrigation scheduling.

*5.3.2.4 Irrigation for qat*

Groundwater abstraction in effect is subsidized and thus may be considered relatively cheap. Yet, many qat farmers in the Sana'a area use expensive pipes and plastic tubes, rather than earthen channels, to convey groundwater from the wells to the various and often distant plots. The reason behind these investments is not to save water as such, but driven usually by the limited supply (many shareholders in a single well). Often, groundwater is pumped from the well to an elevated small pond, from where water is released through a distribution network that conveys water to each plot. It is common to find several kilometers of connecting pipes to convey groundwater to fields located in areas with limited availability of groundwater. Farmers justify the extensive use of pipes as to minimize conveyance losses associated with the use of earthen channels. Within a plot, water is applied between the qat rows, which on average are around one meter wide. Rows are usually divided into sections with the higher sections watered first. Within a plot, water is either applied to one row and then channeled to the next row, or plastic hoses are used for the same purpose. Qat reacts strongly to watering. After several good rains or irrigations, qat shoots out and thus can be pruned. This causes the over-supply of qat during the rainy summer months, while the opposite holds true during the winter months.

The survey results show that the water application for qat varied widely and averaged around 1.25 m/a (Figure 5.9). Eighty two percent of the surveyed qat farms received less than 1.8 m of irrigation water per year. While the overall average number of irrigation sessions for the qat plots of the pilot farm was around 12 per season, ninety percent of the surveyed farmers limited water application to only 3-12 irrigation sessions per season.

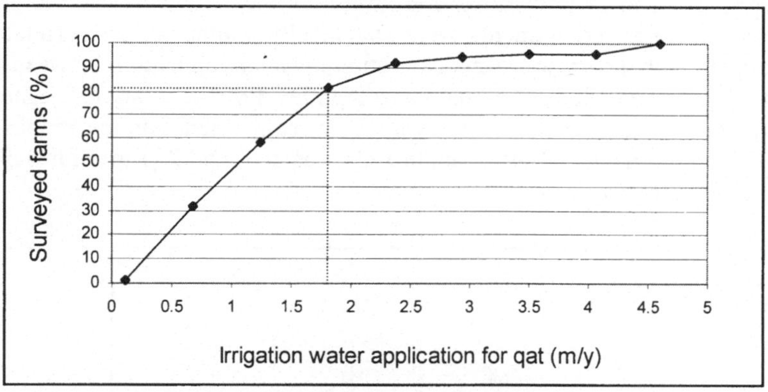


Figure 5.9 Cumulative percentage of irrigation water application for surveyed qat farms in the Sana'a basin (n=74).

Compared with the survey results, water application data from the pilot farm do not show large variations. With the exception of Q1, which was only planted in 1994 and thus was not harvested during the '96 and '97 seasons, qat plots show a water application range of 1.1-2.0 m/season (Figure 5.10). Realizing that irrigation water application is influenced by climatic and market factors, this variation in water use may be expected. The variation in water application becomes narrower (1.3-1.6 m) when expressed as an average of all qat plots within the pilot farm. The overall average water application over the three seasons of the study was 1.5 m/a.

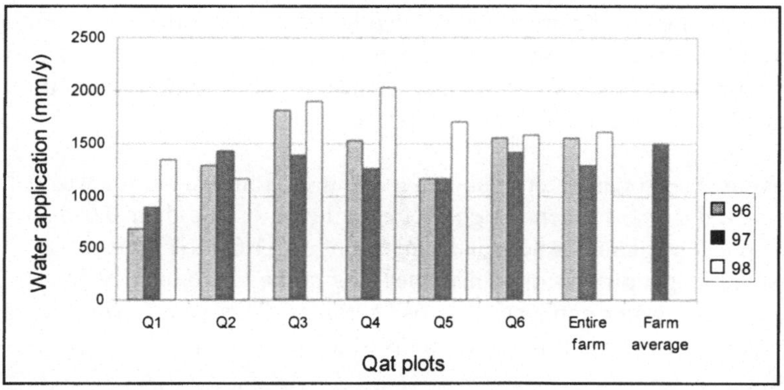


Figure 5.10 Irrigation water application for qat plots in the pilot farm for the irrigation seasons '96, '97 and '98.

The calculated theoretical values of the monthly irrigation water requirement are compared with the actual application (Figure 5.11). Unlike the actual water application, the calculated values are more distributed throughout the year. Although the theoretical irrigation water requirement totals 1,318 mm/a[69], the actual total average water application to qat was 1,560 mm, 1,298 mm and 1,611 mm in the '96, '97 and '98 seasons, respectively. During the '96 season, water was purchased from neighboring farmers to irrigate qat and thus compensate for the low yield of their own well (for a discussion on water markets in the Sana'a area refer to Chapters 2 and 6).

[69] The monthly irrigation water requierment for qat was obtained from HWC (1992a).

When the three data sets of monthly water application for the '97, '98 seasons and the theoretical monthly water requirements were statistically compared, no correlation was found. Coupled with a narrow seasonal variation in water application between the three data sets, it can be concluded that qat has a high level of flexibility in irrigation scheduling, where for example the theoretical water requirement for the month of April is around 70 mm, the actual water application was zero and 235 mm for the'97 and '98 seasons, respectively.

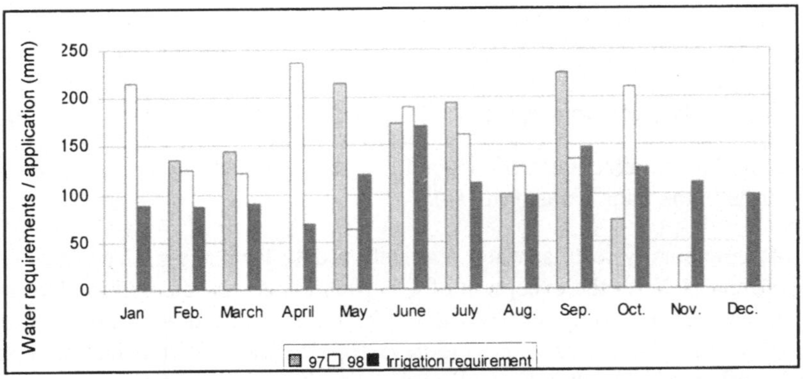


Figure 5.11 Monthly application of irrigation water for qat in relation to theoretical irrigation water requirements (pilot farm).

*5.3.2.5 Irrigation of grapes*

A typical grape plot in the Sana'a basin area consists of several planted rows. Irrigation water is applied to the grape rows, ranging 2-5 m in width. Unplanted land between the rows (1-10 meters in width) is sometimes watered to enhance the growth of the root system of the grape trees, and, depending on water availability, is sometimes planted with qat, cereals or fruit trees.

Although the average seasonal water application obtained from the survey was 0.8 m, 50% of the surveyed grape farmers irrigated to the level of less than 0.6 m, while another 40% applied irrigation water in the range of 0.6-1.6 m (Figure 5.12). The number of seasonal irrigation sessions for grapes was in the range of 3-10, with 62% of the farmers limiting water application to 5-7 sessions. This is compared with an average of 9 irrigation sessions for the grape plots of the pilot farm.

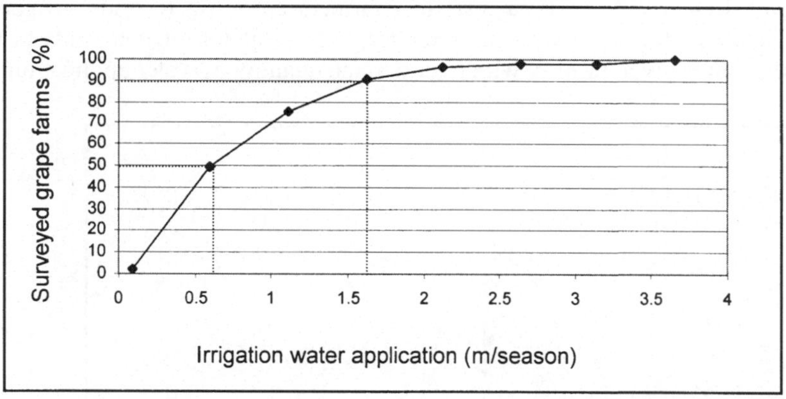


Figure 5.12 Cumulative percentage of irrigation water application for surveyed grape farms in the Sana'a basin (n=54).

Like for qat, a smaller range of irrigation water application was used (0.46-1.76 m/a) for the various plots of the pilot farm than was deduced from the survey. Figure 5.13 shows the overall average range of water application for the entire farm to be 0.64-1.3 m/a. Moreover, the overall average water application for the three seasons was 0.95 m/a. The large increase in water use during the '97 season may be attributed to greater accessibility to water after the rehabilitation of the borehole with larger capacity equipment. Although water application was eventually reduced during the '98 season, it remained considerably higher than that of the '96 season (on average 40% higher). This suggests that the farmer had been facing shortage in irrigation water with the old equipment. Unlike for qat, no additional water was purchased for grape irrigation during any of the three seasons, which finds its rationale in the larger needed water quantities and the expected lower financial returns compared to qat (Chapter 6).

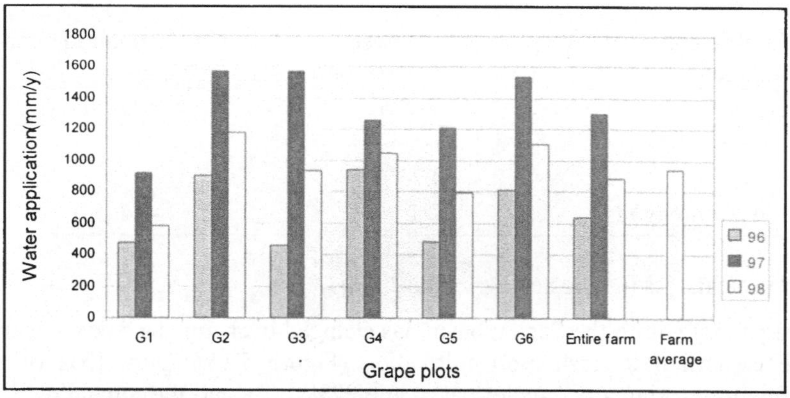


Figure 5.13 Irrigation water application for grape plots in the pilot farm for the irrigation seasons '96, '97 and '98.

While the theoretically calculated monthly irrigation water requirement for grapes emphasizes water use during the second half of the year, farmers, in reality, concentrate water application during the first half of the year (Figure 5.14). Unlike for qat, the total yearly water application for grapes, 0.64m, 1.3m and 0.89m for the '96, '97 and '98 seasons, respectively, was much lower than the yearly irrigation water requirement of 1.64 m/a[70]. This suggests that farmers are close to under-irrigating grapes, which, in combination with the recorded over-irrigation of qat, leads to the conclusion that qat is given priority with regard to the quantity of water application.

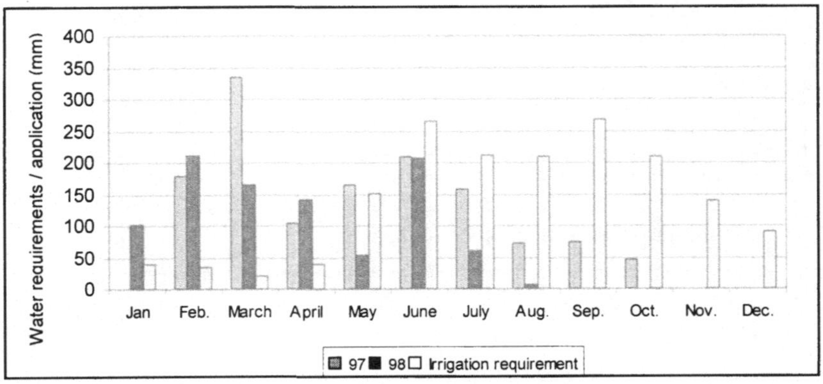


Figure 5.14 Average monthly application of irrigation water for grapes in relation to irrigation water requirement.

Table 5.4 shows the results of a cross-correlation analysis of the monthly water application in '97, '98 and the theoretical irrigation water requirement. A low and even negative correlation coefficient exists between the calculated monthly irrigation water requirement and the actual monthly water application during the '97 and '98 irrigation seasons, but a high correlation coefficient of 0.7 exists between the recorded monthly water application of the '97 and '98 seasons. Unlike for qat, the reasonably high correlation between the two seasons indicates that farmers follow a similar irrigation pattern every season indicating that grapes are given priority in irrigation scheduling, and confirming the sensitive nature of grapes towards irrigation timing (Doorenbos and Kassam, 1979).

Table 5.4 Coefficients of correlation between calculated irrigation water requirement and field application of irrigation water for grapes.

| | '97 season | '98 season |
|---|---|---|
| '97 season | - | - |
| '98 season | 0.7 | - |
| Calc. irrigation requirements | -0.1 | -0.4 |

### 5.3.3 Land use changes in irrigated agriculture

The land use pattern within the Sana'a basin has changed over time as a consequence of the large expansion in cash crop cultivation (Figure 5.15). Only 13% of the surveyed farms were originally[71] under qat, while 32%, 20% and the remaining 35% have been converted from uncultivated agricultural land, rain-fed cereal cultivation and rain catchment land, respectively. These percentages reflect the criteria requirements for land selection for qat. It is widely known among the local farmers that qat does not require highly fertile soil nor does it need a thick soil layer. The need for protection from frost and cold winds makes land, previously designated for rain

[70] The monthly irrigation water requirements for grapes was obtained from HWC (1992b).
[71] Prior to reliable access to groundwater (until the early 1970s).

harvesting, appropriate for qat. These arguments also explain the relatively low conversion from land designated previously for rain-fed cereals to irrigated qat cultivation.

At the same time, grape cultivation expanded from the original 13% of the surveyed farms; 56% and the remaining 31% have been converted from uncultivated agricultural land and rain-fed cereal cultivation, respectively. Unlike for qat, soil requirement for grape cultivation restricts any conversion from rain catchment areas. Although the results do not provide accurate values of the total area under each crop within the basin area, it describes the trend of change.

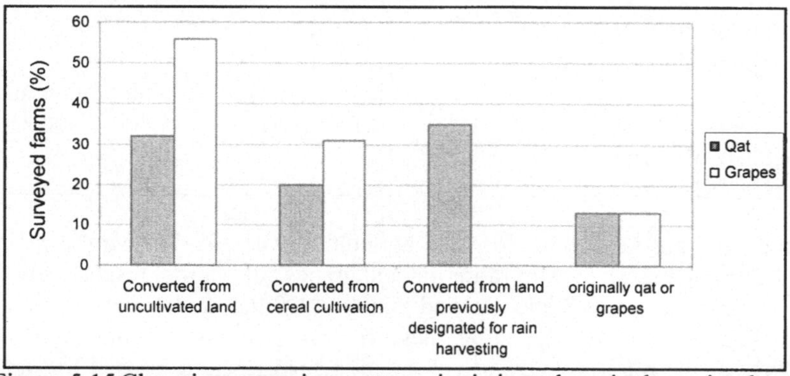


Figure 5.15 Changing cropping pattern in irrigated agriculture in the Sana'a basin ($n_{qat}$=74; $n_{grapes}$=54).

In an effort to estimate the area under qat and grapes within the Sana'a basin, Table 5.5 specifies the type and source of the required data.

Table 5.5 Type and source of needed data to determine the total groundwater abstraction and the relative area under qat and grape cultivation.

| **Type of data** | **Data source** |
|---|---|
| Average well yield | Field survey, verified by pilot farm |
| Operation hours of wells | Field survey and pilot farm |
| Water application for each crop | Field survey and pilot farm |
| Total number of wells within the basin area | Well inventory carried out by the SAWAS project. An additional 10-25% more wells are added to the inventory records to compensate for uncounted and new wells |
| Share of each crop in the overall water use | HWC, 1992a |

The approximate groundwater use for irrigated agriculture, and the area under qat and grapes within the Sana'a basin in 1996, are shown in Table 5.6.

Table 5.6 Total groundwater use for agriculture and the area under qat and grapes in the Sana'a basin area in 1996.

| Item | | Value |
|---|---|---|
| Average well yield | (a) | 4.4 l/s (15.84 $m^3$/h) |
| Operational hours of a typical well (hours/day * days/month * months/season) | (b) | (12*30*8) = 2,880 h/a |
| Seasonal abstraction of a typical well | (c)=(a)*(b) | 45,620 $m^3$/a |
| Number of wells within the basin area | (d) | 4,500-5,500 |
| Total abstraction within the basin area | (e)=(c)*(d) | 205-250 $Mm^3$/a |
| Share in water use for qat (40%) | (f) | 82-100 $Mm^3$/a |
| Share in water use for grapes (25%) | (g) | 51-63 $Mm^3$/a |
| Field water application (qat) | (h) | 12,500*-15,000** $m^3$/ha |
| Field water application (grapes) | (i) | 8,000*-9,500** $m^3$/ha |
| Area under qat | (j) = (f)/(h) | 5,500-8,000 ha |
| Area under grapes | (k) = (g)/(i) | 5,400-7,900 ha |

* Average of the field survey.
** Average of the pilot farm.

Total groundwater abstraction for irrigated agriculture was 205-250 $Mm^3$/a in 1996, with 82-100 $Mm^3$/a and 51-63 $Mm^3$/a for qat and grapes cultivation, respectively, on a total area of around 5,500-8,000 ha and 5,400-7,900 ha under qat and grapes, respectively (Table 5.6). In calculating these figures, two assumptions were used: 40% and 25% of the total groundwater abstraction for irrigation is used for qat and grapes[72], respectively; and an estimated 4,500-5,500 wells are used for agricultural irrigation within the basin area. However, these two assumptions need to be further verified. The first assumption indicates that the ratio of water use for grape irrigation is around 62% of that for qat, which is in close agreement with the 57% ratio found in the surveyed farms that grew both qat and grapes. As for the second assumption, it is based on the well inventory carried out by the SAWAS project in 1993. The inventory results showed the total number of visited wells in the basin area to be around 4,500 wells. Realizing that many wells may not have been inventoried, the actual number of wells within the basin area was then estimated at 5,000 with a variation of ±500 (NWSA, 1996).

## 5.4 Conclusions and recommendations

The large scale groundwater development during the 1980s, in combination with extensive sharing of boreholes among farmers, has provided a large portion of the basin population with a reliable water source, thus encouraging rapid expansion of irrigated agriculture. The high involvement of the basin population in well ownership and thus in irrigated agriculture, increases the difficulty to manage and influence irrigation, especially when community participation cannot be guaranteed. Extensive well sharing enhances access to a reliable irrigation source, thus increasing equity, and productivity; the shift to more valuable crops increased also the value of groundwater to the farmers. However, this has only been possible at the expense of the resource sustainability.

[72] According to HWC (1992), Water consumption of qat, grapes and vegetables within the Sana'a basin were estimated at 40%, 25%, and 35%, respectively.

Farmers in the basin area behave rationally, but are strongly risk-avert in crop choice, agronomy and irrigation habits. One way for the government to promote lower water use is to help farmers reduce the risk, for example by better agricultural extension, introduction of water saving technology, and information assistance to shift to other crops. With limited future potential for irrigated agriculture in the basin area, the government should eventually seek to invest in an alternative economic base to replace the current fragile water based economy.

In the Sana'a basin, the very large number of shareholders in wells on one hand complicates any effort to regulate abstraction. On the other hand, as this network of shareholding coincides to some extent with tribal structures, this may open avenues to improve awareness and facilitate collective action to reduce water abstraction.

Irrigation water application for qat in the Sana'a basin was higher than the theoretical irrigation requirement, reflecting an effort by farmers to reduce the risk of lower yield. Water use and thus revenues from qat are an important component in the budget of a rural household, summarized by a farmer saying "I may spend less on food and water for me and my family, but will not reduce water to qat. It is our main source of income". Despite the fact that most of the surveyed farmers considered grape cultivation to be profitable, water application was consistently found to be less than the theoretical water requirement. Most farmers view grape cultivation to be more water and labor demanding than qat. It can be concluded that while for agronomic reasons grape was given priority in irrigation scheduling, it is under-irrigated. On the other hand, although qat was given more attention in the quantity of irrigation water often resulting in over-irrigation, it allowed a high level of flexibility in irrigation scheduling. Farmers would purchase additional irrigation water to save their qat crop, but would not do the same for grapes.

Water availability has changed the cropping pattern in the Sana'a basin area, converting large areas of traditional rain harvesting land into qat cultivation, and uncultivated land and land under cereals into grape production.

Water application for qat and to a lesser extent for grapes showed wide fluctuations, which may be the consequence of a failing agricultural extension service. It was observed during the survey that farmers do not trust the government and its officials. Although there was no effective extension program in the surveyed region, in areas where they are established, most often qat farmers are excluded from any services. In an effort to enhance the field level irrigation efficiency, the government (with World Bank funding) has initiated the "Land and Water Conservation" project. Although the project provides farmers the opportunity to buy highly subsidized conveyance pipes, qat farmers are excluded since they are viewed to be financially capable of purchasing pipes at regular market prices. Although this policy may seem to promote equity and agricultural (mainly cereal) production, it should be kept in mind that irrigated agriculture is dominated by cash crops, of which qat is the leading crop. In other words, qat is sufficiently profitable to remain a very powe Average of the pilot farm.rful incentive for farmers to continue the existing cropping pattern. It is true that qat farmers may not be in need for subsidized water conveyance pipes, but a national policy to conserve water resources should concentrate on ways to enhance irrigation efficiency to qat through research, awareness programs and strong extension services that gives priority to qat.

## 5.5 References


Central Bureau of Statistics (1993) Agricultural survey. Ministry of Planning and Development, Sana'a, Yemen

Central Bureau of Statistics (1995) Results of the General Population Census. Ministry of Planning and Development, Sana'a, Yemen.

Doorenbos J. and Kassam A. (1979) *Yield response to water*. FAO Irrigation and Drainage paper 33, Rome, Italy.

Doorenbos J. and Pruitt W. (1984) *Crop Water Requirements*. FAO Irrigation and Drainage paper 24, Rome, Italy.

Easter K., Rosegrant M. and Dinar A. (1999) Formal and Informal Markets for Water: Institutions, Performance and Constraints. *The World Bank Research Observor*. **14**(1), 99-116.

Fujita K. and Hossain F. (1995) Role of the groundwater markets in agricultural development and income distribution: A case study in a north-west Bangladesh village. *The Developing Economics*. **33**(4), 442-463.

HWC (1992a) Water Resources management Options in Sana'a Basin. Final Report Vol. 9, Technical Secretariat of the High Water Council, Sana'a, Yemen.

HWC (1992b) Regional Water Requirements for different water-consuming sectors. Final Report Vol. 5, Technical Secretariat of the High Water Council, Sana'a, Yemen.

Mohindra M. (1981) Improvement of water use efficiency in Vineyards of Yemen Arab Republic. Agriculture Research Service, Ministry of Agriculture, Sana'a, Yemen.

NWSA (1996) Evaluation of the effects of groundwater use on groundwater availability in the Sana'a basin. Technical Report No.5, vol. 5, SAWAS-3, National Water and Sanitation Authority, Sana'a, Yemen.

Saleth R. (1996) *Water Institutions in India: Economics, Law, and Policy*. Commonwealth Publishers.

Shah T. (1993) *Groundwater Markets and Irrigation Development: Political Economy and Practical Policy*. Oxford University Press.

Shah T. (1997) Integrating Water Markets in Sustainable Water Resources Management. *World Bank Research Observer*.

Ting-Tiang (1998) Crop requirement tables. AREA/FAO. Ministry of Agriculture and Irrigation, Sana'a, Yemen.

# Chapter 6

## Financial Incentives for Groundwater Use in Irrigated Agriculture in the Sana'a Basin

### 6.1 Introduction

Groundwater development for domestic use as well as for agriculture started in the Sana'a basin area in the early 1970s. After a decade of increasing private well development, a drop in groundwater levels in most of the basin area became an evidence of heavy groundwater mining. The government recognized the fossil and vulnerable nature of the aquifer early in the 1970s and passed a groundwater protection law that introduced a licensing system for all underground drilling within designated protection zones in the Sana'a basin (Ministry of Legal affairs, 1973). The lack of a suitable and effective institutional setup, not only to monitor and enforce the newly introduced permit system, but also to respond to the increasing water demand resulted in an accelerated depletion rate reaching an annual 2-6 meter drop in water levels in most regions of the basin. This drop has been caused mainly by the rapid expansion in private irrigated agriculture increasing the number of supplying wells to more than 5,000 in the mid 1990s (NWSA, 1996). Some reports have predicted that if groundwater abstraction continues at the current level, the aquifer will run dry in many locations by the year 2010 (Foppen, 1996).

Realizing that groundwater is the exclusive water source in the area, and given the high financial, economic, legal, social and technical constraints of water import from other regions (Chapter 2), it becomes of great importance for the government to regulate this valuable resource as to maximize the benefits from its use in a sustainable manner. Nevertheless, the government has adopted a policy that in effect promotes agricultural production by subsidizing energy (mainly diesel) used for groundwater pumping and setting low import duties on water extraction equipment (pumps, generators and pipes). Therefore, the aim of the government to regulate abstraction has been overridden by the conflicting aim to increase agricultural production. In addition, the subsidization of imported grains has driven national grain cultivation out of the market thus shifting the cropping pattern to more water demanding cash crops, particularly qat, grapes and vegetables. Subsidized water in combination with high profits to farmers from cash crops are considered to be the main driving force behind the rush to expand irrigated agriculture.

National policy with regard to the management of groundwater resources is not consistent, particularly where it concerns the Sana'a basin. This inconsistency is mainly attributed to the lack of coordination between the water sub-sectors, where conservation of groundwater (through a permit system) is considered a domestic water supply concern, while promotion of water use for agriculture (fuel subsidies and low import duties) is the concern of the Ministry of Agriculture and Irrigation

(MAI)[73]. This conflict in interests was very apparent when the responsibility to monitor and issue the drilling permits within the Sana'a basin area (allocation of the resource) was taken from the National Water and Sanitation Authority (NWSA)[74] and handed to the MAI. Thus, MAI became a water manager (regulator) and a water use promoter at the same time. The establishment of the National Water Resources Authority (NWRA) in 1995 is a step in the direction to eliminate the conflict, confining the role of MAI and NWSA to being only water users. Although the right institutional setup is essential for proper groundwater management, the situation in Sana'a is further complicated by (1) the low capacity of the government to monitor and enforce any measures to control drilling and abstraction, especially since groundwater has been exclusively developed by private investment, (2) the lack of a legal framework to support a policy to allocate and reallocate groundwater to ensure optimum productivity and sustainability, especially given the customary system of water ownership and water use rights, (3) the local social and political power structure, with the strong role of the tribes weakening the effectiveness of the central government, and (4) the strong current financial incentives to farmers in the form of subsidies and high profits from cash crops.

Given the political-economical constraints to the implementation of any command-and-control regulatory measures to manage groundwater, it becomes essential to examine the potential role of financial and economic regulatory measures for water demand reduction. To achieve this, the current cost and value of water use both to the domestic sector and agricultural users need to be determined, as these form a major part of the incentive system that guides the behavior of the water users.

The cost and value of water use in the domestic sector comprising all urban water supply alternatives, the capacity and willingness of consumers to pay for water supply, and identification of the constraints to water supply augmentation, were examined in Chapter 2. This chapter deals with (1) the evaluation of the cost and profit generating value of groundwater use for qat and grape cultivation in the Sana'a basin, and (2) the effects of higher input costs on groundwater use for qat and grape cultivation. The relative components of income, profit, profitability per unit of land and water, and cost-benefit ratios for qat and grape cultivation will be clarified. The effects of the energy cost component for groundwater pumping in the overall farm-level micro-economics of qat and grape cultivation will be emphasized. This understanding will address the critical question whether a rise in diesel price would lead to a reduction in groundwater irrigation for cash crop (qat and grapes) in the Sana'a basin.

## 6.2 Methodology

Information regarding qat cultivation, especially of agronomic and financial nature, is scarce. This has been mainly the result of official policy to exclude qat cultivation from any extension and agricultural programs. Grape cultivation, although not excluded from extension programs, is not given much attention, not only in the Sana'a basin but also in other grape growing areas such as Amran. Therefore, this research aims to collect basic detailed data on cultivation within the study area.

[73] Formally named Ministry of Agriculture and Water Resources.
[74] NWSA is responsible for urban water supply and sanitation services in all Yemeni cities and major towns.

The methodology was based on two complementary techniques. First close long-term and detailed monitoring of all agronomic, irrigation, and financial activities related to qat and grape cultivation in a farm representative for the Sana'a basin (the "pilot" farm). In addition to general observations, daily records were taken of all financial aspects related to the crop cultivation, which included (1) the purchase of diesel, lubricating oil, pesticides, fertilizer, clay, etc., (2) the payments for maintenance, and (3) all payments related to labor work within the farm. In this farm, the sons of the farm owner carried out most of the labor work, which was made transparent by converting this to actual costs based on the local labor compensation rate. In addition to the input costs, all sales were recorded, which for grapes included the harvest (number of boxes) and the selling price per box. The sales of qat were expressed per plot due to the difficulty in measuring its yield. The pilot farm was monitored for three consecutive '96, '97 and '98 seasons. This duration allowed to (1) investigate the effect of improved accessibility to irrigation water after the replacement of the aging small diesel generator and pump with higher capacity equipment in the beginning of the '97 season, and (2) to examine the effect of the rise in diesel price on farming and water use at the end of 1997.

The second technique consisted of a broader field survey in the Sana'a basin to confirm the data obtained from the pilot farm and test the representativeness of the pilot farm data. Sixty seven wells and 79 farms were surveyed, of which 49 farms had a combination of qat and grape plots and 25 and 5 farms had only qat or grape plots, respectively. Although the field survey was conducted in June 97, data were collected as to represent the '96 season. Location and selection criteria of the sampled farms are discussed in details in Chapter 5.

Data acquired from the pilot farm were more accurate than those of the field survey since the survey respondents had no written records and relied mostly on recollection and sometimes even on estimates. However, given the wide geographical coverage of the survey including almost the entire northern part of the basin area, it provided a wide range of perspectives on views and perceptions towards issues such as water ownership, water use rights, conflict resolution, and groundwater management. During the survey it was observed that some respondents were reluctant to give out information regarding the finances of farming, especially those related to sales and profits, so the questionnaire was modified to request information on only one typical average plot of each crop instead of on the entire farm. By doing so, farmers could be more precise and felt that they were not exposing their financial status since the discussion focused only on a small part of their property. To obtain the actual irrigation water application, the yields of the supplying wells of the sampled farms were measured using a stopwatch and a marked 50-liter bucket. The setup of the questionnaire, reliability of the collected data and the time spent at each farm are discussed in detail in Chapter 5.

## 6.3 Results and discussions

### 6.3.1 Costs of qat and grape cultivation in the Sana'a basin

#### *6.3.1.1 Total costs*

In general, there are three different cost components in irrigated agriculture, namely (1) water, (2) labor and (3) other. Other costs include the purchase of fertilizer, pesticides and clay. Pesticide use for qat and grape cultivation is believed to be widely practiced although in many cases this is denied by farmers. Although farmers believe that dusting[75] gives a similar results to pesticides use, it is likely to be ineffective during the rainy season. The use of manure as a local fertilizer is widely practiced and in recent years this is being increasingly supplemented with chemical fertilizer.

Even though in a water short region like Sana'a water should the most critical input to irrigated agriculture, its quasi unrestricted accessibility reduces drastically its relative financial importance to the farmer. Figures 6.1 and 6.2 show the different cost components for qat and grape cultivation, respectively.

While the water-related cost component (including capital, operation and maintenance) represented 66-88% of the total costs for qat cultivation on the pilot farm, this was around 67% in the field survey. Total water-related cost was 18.0 US¢/m$^2$ in '96 and '98, and 10.6 in '97 on the pilot farm, and 12.6 on average in the 1996 survey. The higher contribution of water in the overall costs in the pilot farm in '96 as compared to '97 as well as in the surveyed farms may be considered exceptional and can be partially attributed to higher maintenance costs of the aging diesel generator and pump, which were replaced before the start of the '97 season. While the total annual cost to cultivate a m$^2$ of the pilot farm with qat in '98 was 24.4 US¢, the 1996 survey results and the pilot farm in the '96 season posted a lower seasonal cost of 19-20.5 US¢. The high costs in '98 can be explained by the 67% rise in diesel prices and a reported comparatively higher water application thus increasing the operational costs to twice that of the '97 season.

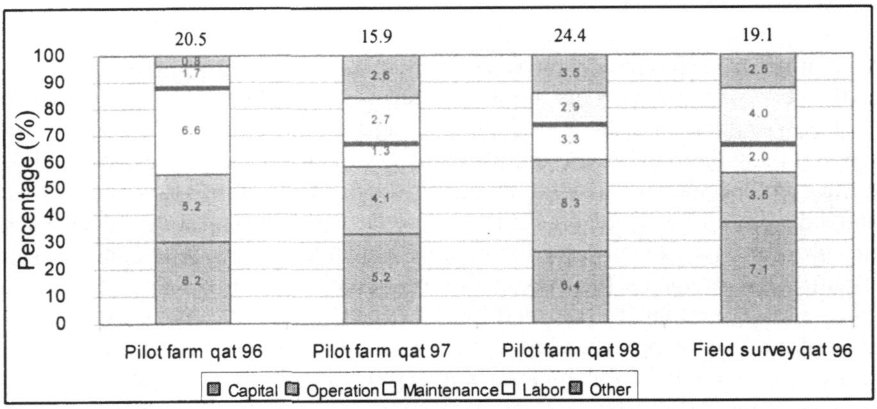


Figure 6.1 The different cost components for qat cultivation in the Sana'a basin. Values in bars are in US¢ per m$^2$. $N_{pilot\ farm}$ = 6 plots, $N_{survey}$ = 72 farms. The total water-related share is the sum of the first three components (capital, operation and maintenance).

As for grape cultivation, the water-related cost component represented 58-68% for the pilot farm and an average of 62% for the field survey, which are very consistent with each other (Figure 6.2). The total water-related cost was 7.4 US¢/m$^2$ in '96, 10.6 in '97 and 9.9 in '98 on the pilot farm and 7.7 on average in the 1996 survey. Crop water application follow different pattern for qat and grapes, which explain the differences

[75] Dusting is the spraying of trees with fine clay, which coats the leaves with a thin layer of dust thus protecting them. This activity is usually performed manually.

in the seasonal evolution between both crops. The higher share of water-related costs for the pilot farm in '97 was caused by a reported higher water application as compared to '96 and '98. A 60% average share of water-related costs seems reasonable for grape cultivation. The total cost to cultivate a m$^2$ of grape for an entire season was US¢ 12.7-16.7 for the pilot farm and averaged US¢ 12.3 in the field survey. Again, both values are consistent.

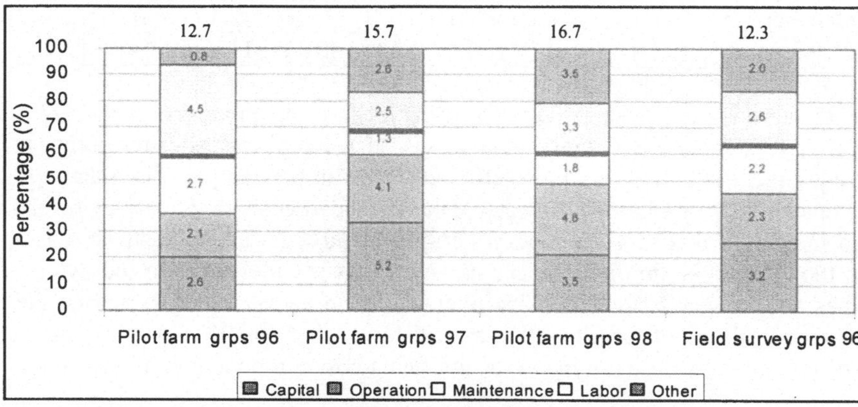


Figure 6.2 The different cost components for grape cultivation in the Sana'a basin. Values in bars are in US¢ per m$^2$. $N_{pilot\ farm}$ = 6 plots, $N_{survey}$ = 53 farms. The total water share is the sum of the first three components (capital, operation and maintenance).

The total annual costs of US¢ 20/m$^2$ and US¢ 15/m$^2$ are appropriate typical averages for qat and grape cultivation, respectively, of which 70% and 60% are water-related, respectively. It should also be noted that these costs exclude the capital cost of the land and the initial investment of plantation.

#### *6.3.1.2 Specification of volume–related costs*

It was difficult to obtain reliable field data for the capital cost of the monitored and surveyed wells due to the fact that those wells were drilled during the 70s and 80s and owners do not have accurate recollection of their capital costs. Therefore, estimates were made based on the current market prices and costs for equipment and drilling (Chapter 2).

Maintenance in '96 constituted a higher share in the pilot farm than in the '97 and '98 seasons, which was caused by the aging equipment (Table 6.1). The maintenance share in the overall water volume-related costs dropped from 37% in '96 to only 12% in '97 after the purchase and installation of newer and larger capacity equipment. The consequent reduction in the use of lubricating oil in '97 contributed to a further decrease in the overall operational unit cost. The higher share of fuel in the '98 season reflected the 67% rise in the price of diesel at the end of '97, resulting in an increase of the fuel share from around 28% in '97 to nearly 34% in '98. The results of the field survey show a trend similar to that of the pilot farm in general and particularly to the '96 season data.

Table 6.1 Capital, operation and maintenance costs of extracted groundwater in Sana'a (US¢/$m^3$).

| **Cost type** | | **Pilot farm '96** | **Pilot farm '97** | **Pilot farm '98** | **Field survey '96** |
|---|---|---|---|---|---|
| Capital | | 4 (34%) | 4 (50%) | 4 (36%) | 4 (32%) |
| Operation | Fuel (diesel) | 2.2 (19%) | 2.3 (28%) | 3.8 (34%) | 2.7 (22%) |
| | Lubricating oil | 1.2 (10%) | 0.8 (10%) | 1.3 (11%) | 2.0 (16%) |
| Maintenance | | 4.3 (37%) | 1.0 (12%) | 2.1 (19%) | 3.7 (30%) |
| Total | | 11.7 (100%) | 8.1 (100%) | 11.2 (100%) | 12.4 (100%) |

The '97 season was a "cheap" year for the pilot farm. Comparison of the '97 season with the '96 average field survey data suggest that the pilot farm after purchasing its new equipment is among the more efficient farms in the sample. The relatively high maintenance share in the overall water volume-related costs in the survey results may be an indication that most extraction equipment throughout the basin area is aging. Even though the rise in diesel prices has increased the fuel share in the overall cost, this has not noticeably increased the overall cost in actual monetary terms; the total water cost in the pilot farm in '98 was still lower than in '96. However, the capital cost of groundwater abstraction was a significant cost component in all seasons.

### 6.3.2 Energy in groundwater extraction

#### *6.3.2.1 Background*

Mining of fossil groundwater usually has two main consequences, namely, (1) higher costs to the present generation to overcome the continuous decline in groundwater level and (2) higher costs to future generations to develop more expensive alternative water resources. Lowering of the water table implies higher pumping head, gradual reduction in well yield and additional capital costs to lower the pump and deepen or replace the well in case of total depletion.

In water short regions (like the Sana'a basin), where energy is subsidized, a rise in energy prices may be considered as an option to better reflect the true opportunity value of groundwater and hence reduce the rate of its depletion. In Yemen, the governmental policy to subsidize diesel (the main energy source for groundwater pumping) is a direct policy to promote agricultural production. Nevertheless, the government, through the National Economic Reform Program, has in recent years started to gradually lower subsidies on diesel raising its price more than threefold in two consecutive steps in 1995 and 1997. Despite this rise, diesel prices are still low because of the simultaneous devaluation of the local currency.

#### *6.3.2.2 International experience*

Previous research on the effects of rising energy prices on groundwater use is inconclusive. While some studies indicate a strong direct correlation between rising energy prices and water use (Heady et al., 1973; Shumway, 1973; Howitt et al., 1980), others have concluded that other factors play a more significant role than energy prices in determining the level of water use in agriculture (Hexem and Heady, 1978; Kelly and Ayer, 1982). These differences may reflect the assumptions employed by the different statistical and programming models. Also a correlation may occur only if the water cost is the primary variable affecting water use. In many regions, the cost of water is comparatively small and thus other factors are likely to influence groundwater use more strongly. According to Carr (1977), these factors include (1) relative prices of farm output compared to the cost of production, (2) national agricultural policies, (3) demand of output products, which is partially a function of taste and preference of consumers, and (4) changing technology. In Arizona, USA a decline in agricultural production and thus lower groundwater use triggered by rising energy prices had been projected, however a study found the actual production and water use to have increased by 8-46% and 10-41%, respectively. It was concluded that even though energy rates had increased, the gross return from crops had increased at an even higher rate, hence countervailing the effects of higher abstraction cost (Carr, 1977).

One of the reasons for the change in attitudes of Texas farmers towards groundwater conservation was attributed to sharp rise in energy prices in the late 1970s (Beaumont, 1985). With increasing draw down and higher energy prices, farmers were expected to voluntary reduce their water use, not for the sake of conservation, but rather because it became unprofitable to continue pumping large quantities of groundwater. A survey of 1927 farmers from 16 western states of the USA, lead Beaumont (1985) to conclude that a cost increase of 25% would trigger a 7% reduction in groundwater pumping. Nevertheless, this reduction in demand was difficult to predict since beside energy prices, other factors such as crop prices and technological improvements strongly influenced water use in irrigated agriculture.

Concentrating only on the additional cost caused by the drop in groundwater level, based on a cost data analysis of changes on irrigation costs in Oklohoma, USA, in 1975 and 1982, Sloggett and Mapp (1984) suggested that although the cost effect of declining groundwater level is strongly related to the local circumstances, generally in areas affected by groundwater mining, declining water levels have not been a major cause of increased overall costs. Nevertheless, it was concluded that a substantial increase in fuel costs and interest rate have contributed to a reduction in the profitability of irrigated agriculture. The higher depletion rate in Sana'a (2-6 vs. 0.3-0.8 m/a) as compared to that in Oklahoma, USA, suggests greater water demand reduction effects from higher energy prices. Nevertheless, the situation in Sana'a differs in the type of cultivated crops with their high demand and thus high profits, reducing the effects of higher energy costs.

In a study aimed at examining the effects of the energy (electricity) rate structure in groundwater use for irrigated agriculture, it was suggested that although the rate of energy use can theoretically affect the aquifer life, the relative crop prices were found to be a more influential incentive in decisions to use water (Gardner and Young, 1984). While these authors suggest that setting energy prices at the marginal cost create an incentive for conservative use of groundwater, Shah (1997) argues that in areas under severe water stress, the social value of water may be higher than that incurred by the marginal cost of energy and consequently even marginal cost pricing of energy may not create a strong enough disincentive for groundwater use. Both Sloggett and Mapp (1984) and Shah (1997) suggest that incorporating the scarcity value of groundwater (the present value cost of the sacrifices imposed on future users by consumption at present) might better be handled through separate policies other than energy pricing, recommending measures such as pumping restrictions, well spacing, severance tax, or the introduction of taxes on energy, drilling or pumping.

#### *6.3.2.3 Energy use for groundwater abstraction in Sana'a*

In the Sana'a basin the dominant energy source for groundwater abstraction is diesel. Although electric pumps are available and some have been installed, their expansion has been hindered by the following factors:

- The high capital and maintenance costs.
- The unreliable electricity service, which results in frequent power cut-offs and in some cases damage to the transformers.
- Installation of electric pumps depends on the coverage of electrification, which in the 1970s was not available in most of the basin area, forcing many farmers to invest in the only alternative of diesel generators.
- Although only two out of 67 surveyed wells were equipped with electric pumps, it was observed that once they are used in a region, their use quickly becomes widespread within that region as electric pumps yield more water than diesel pumps.
- Farmers in some regions have collectively agreed not to install electric pumps in order to minimize externalities (interaction between wells), which may be viewed as an effective community-based measure to ensure equitable access to groundwater.
- Farmers feel less secure with electric pumps realizing that a third party (the power company) also controls operation decisions.

For these reasons and given the shallower groundwater levels in the 1970s, most farmers then preferred to install small (one cylinder-24 hp) diesel generators. These engines became very popular for their low price and simple operation and maintenance. As groundwater levels deepened, some farmers started to shift to more powerful 6-cylinder engines. The survey results show that 6- cylinder diesel engines are installed on 30% of the sampled wells, 45% of them having replaced smaller one-cylinder generators. The power (capacity) of the generators correlated reasonably well with depth to the water level with a correlation coefficient of 0.6. Reasons for not obtaining a higher correlation coefficient may be:

- Some 6-cylinder engines are installed on wells with relatively shallow groundwater level, thus possibly reflecting a high water demand (many shareholders) or anticipated future drop in water levels.
- The widespread sharing of wells raise the difficulty to reach a consensus among all shareholders of a well to upgrade the engine, especially when shareholders dispose of a varying financial capacity. In the pilot farm, for example, share holders discussed and argued about changing the generator for about three years before they took action. This suggests that with continuous decline in groundwater level, eventually all small generators will be replaced by larger capacity engines.

#### *6.3.2.4 Diesel consumption in the Sana'a basin*

Consumption of diesel depends mainly on two factors: engine efficiency and depth to the water level. On the pilot farm diesel consumption was 0.4-0.5 $l/m^3$ of pumped groundwater. The survey showed an overall average of 0.6 $l/m^3$, but this decreased to 0.45 $l/m^3$ if the wells in the area of Al-Gherass[76] were excluded (Figure 6.3).

[76] The survey showed that groundwater in the area of Al-Gerass is at an exceptional depth of 165-210 mbgl with an average of 188 mbgl, whereas the average depth to groundwater in the rest of the basin area is only 138 mbgl.

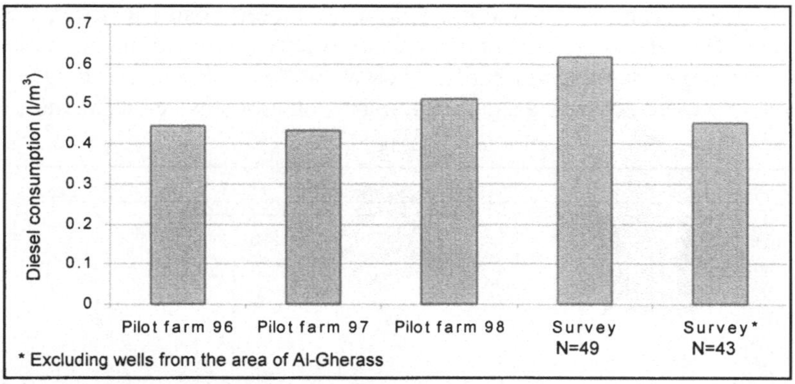


Figure 6.3 Diesel consumed per unit of pumped groundwater ($l_{diesel}/m^3{}_{groundwater}$).

Schiffler (1998) reported that in the highlands of Jordan, diesel pumps need between 0.25 and 1 liter of diesel to pump 1 $m^3$ of water, depending on the depth of the groundwater and the type of the pump. Data from the pilot farm as well as from the surveyed farms suggest that roughly half a liter of diesel is required to pump one $m^3$ of groundwater from an average depth of 138 mbgl. With the subsidy on diesel estimated at 50% of its price, it can be concluded that the government provides farmers with a free quarter liter of diesel for every $m^3$ of pumped groundwater. In monetary terms, this means that farmers indirectly receive 5 YR/$m^3$ of pumped groundwater and thus consume around 91,200-136,900 $m^3$/a of diesel and pose a burden of around YR 1.0-1.3 billion[77]/a on the government from diesel subsidies.

*6.3.2.5 Cost of diesel*

At first hand it may be concluded that the resistance to remove the subsidy on diesel and the high profits from cash crops render diesel price an ineffective regulatory instrument to control water demand. Nonetheless, its availability has been observed in this study to be sufficiently critical that farmers travel great distances and pay 2-3 times the official price during periods of shortages.

Figure 6.4 shows the relative share of diesel in the overall production costs, the water related costs, and the O&M costs of a typical qat-and-grapes farm. The diesel share in the overall costs ranged from 12% in '96 to 20% in '97 and an average of 24% for the surveyed farms. The diesel share in the overall costs increased to 26% in '98 after the 1997 rise in diesel price. A similar trend, although with higher percentages, can be seen for the share of diesel in the water-related and O&M costs with ranges of 19-34% and 29-56%, respectively. The sharp increase in maintenance costs in '98 lowered the diesel share in the O&M costs to slightly below that of the '97 level. This demonstrates that even a 67% rise in diesel price is quickly offset by increases in the cost of other inputs, especially unpredicted maintenance, and likely by shifts in market prices of crops as well. The situation in Sana'a is relatively comparable to the highlands of Jordan, where a 40% increase in diesel prices (from 75 to 105 Fils/liter)[78] was calculated to increase pumping costs by 7.5-30 Fils/$m^3$, which is less than 1% (for

[77] This is based on the survey data of an estimate of 4,500-5,500 wells in the basin (NWSA, 1996), an annual average of 2,880 operational hours per well, and the average well yield of 15.84 $m^3$/h. At the exchange rate of YR 150-160 to the US$ (1999), the economic burden on the government is US$ 6.3-8.7 million.

[78] 1 Jordanian Dinar (1000 Fils) is equivalent to 1.43 US$.

greenhouse vegetables) and 30% (for olives) of the total direct costs (Schiffler, 1998). He further reported that any realistic increase in electricity (a second major source of energy for groundwater pumping) tariffs (below 100%) will simply reduce the farmers' profits, but will not have a significant impact on groundwater abstraction.

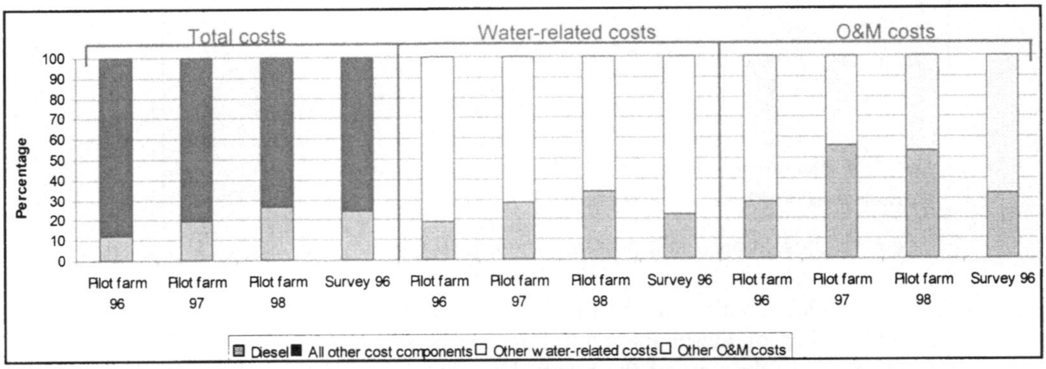


Figure 6.4 Cost of diesel in the total, water–related, and O&M costs of irrigated agriculture for qat-and-grapes farm in the Sana'a basin.

## 6.3.3 Income and profits

### *6.3.3.1 Income*

The survey results showed that, with the exception of a few farmers who obtain extra income from regular work mainly in Sana'a, farmers depend primarily on crop sales for income. Qat sales in the Sana'a basin are conducted either through wholesale or retail marketing. Under the wholesale procedure, a middleman, *mofawed,* buys the right to harvest a number of rows or an entire plot. He is then responsible for the pruning, transport and selling of the harvest to the consumers. During high supply episodes on the qat market (summer months), the middleman may spend several days or even a week to prune a small plot. Qat farmers in the wholesale procedure receive a lump sum price (sometimes in installments), determined by the quantity and quality of the harvest and, equally important, by the anticipated qat supply and demand in the market. Under retail sale on the other hand, farmers themselves prune, bundle, transport and sell the harvest at the marketplace to consumers. Even though in the latter procedure farmers earn more profits, most farmers prefer the first method for its lower risk[79], labor and time requirement. Almost all qat harvested in the Sana'a basin is marketed mainly in Sana'a city and a smaller portion in the surrounding rural communities.

Qat price and farmers' income thus depend on two main variables, namely the harvest quality and the over-or-under-supply compared to demand. The quality of the harvest for a given plot depends on many variables such as soil type, plot location and, most importantly, the origin and type of the planted trees. Hence, qat quality from a particular plot can be considered more or less constant over time.

[79] Since qat is usually sold fresh (chewed the same day it is pruned) any excess unsold quantity loses much of its value if left for the second day. Middlemen are perceived to be better able at assessing the market and thus prune the right quantity.

As for supply and demand of qat on the market, it can be assumed that supply plays a more important role in price setting given that short-term demand is more or less constant[80]. In order to assess the supply of qat, it is important to understand its temporal availability in the market.

Although farmers were found to perceive the qat tree to be so flexible[81] that it can be harvested at any time of the year, the risk of losing the trees during certain weather conditions (a combination of cold wind and dew conditions) limit productive time to the warmer spring and summer months. After several water and pesticides applications, qat trees can be ready for harvest within 2-3 weeks. In the Sana'a basin, qat trees are usually pruned twice a year. Intensity and frequency of water application are keys to prepare the qat trees properly for harvest. After periods of rains most farmers are faced with a ready harvest, resulting in an over-supply of qat on the market and consequently lower prices. Given the large qat market in Sana'a, qat from other regions of the country is also marketed in Sana'a increasing supply further driving down its price especially during the rainy summer months. Therefore, the highest net gains can be obtained from qat plots that lie in sheltered locations so that irrigation and harvest can proceed in the low-supply winter period.

Unlike qat, grapes are reported to be harvested in a distinct season; nonetheless, due to the different varieties, the grape season usually extends from early spring to late summer. Farmers usually harvest (pick) and sell the yield themselves. During periods of high supply in Sana'a, some farmers may have to market their harvest in other cities. It was also observed that some farmers sell their harvest at the farm to middlemen on volumetric basis, and set the price per container (10-15 kg).

Table 6.2 shows the sales of qat and grapes form the pilot and surveyed farms. Income from qat sales increased between '96 and '98, which may be the result of growing and partly unmet demand. The range in income from sales is quite narrow for the different plots of the pilot farm; the larger range in the field survey may be explained by the following two factors:

- A large variation in qat quality causes a wide range of market prices and income to farmers.
- In contrast to the pilot farm, where all sales were recorded accurately, interviewed farmers were in many cases reluctant to give out precise information regarding sales and income. Some farmers tended to not report the exact sales, but rather reported what they considered average or fair for their plot or farm. In some cases reported sales and income are believed to have been exaggerated or underestimated. However, it is believed that the averages and tendencies found are sufficiently accurate.

---

[80] It is true that demand for qat may be growing with population growth, in the short-term i.e. one season, or one year, it may be considered constant.

[81] This flexibility is evident from the saying of an interviewed farmer "A qat tree is like a safe, whenever you need some money, just irrigate and apply some pesticides and there you are, it is ready for harvest".

Table 6.2 Average annual sales of qat and grapes from the pilot farm and surveyed farms ($N_{pilot\ farm}$ = 6 plots; $N_{surveyed\ qat\ farms}$ = 73; $N_{surveyed\ grape\ farms}$ = 49). All values are in US$ per $m^2$ of cultivated land (values between brackets show the range).

| | Pilot farm '96 | Pilot farm '97 | Pilot farm '98 | Field survey '96 |
|---|---|---|---|---|
| Qat | 1.03<br>(0.8-1.2) | 1.21<br>(1.0-1.4) | 1.55<br>(1.0-2.2) | 1.0<br>(0.1-4.1) |
| Grapes | 0.41<br>(0.3-0.6) | 0.63<br>(0.35-1.2) | 0.84<br>(0.5-1.3) | 0.33<br>(0.03-2.3) |

Like qat, income from grapes has also increased in the period of 1996-1998, probably indicating a growing and only partly satisfied demand over the years. Although the above mentioned explanations for the large variation in income from qat sales holds true for grapes as well, grape yield is much more sensitive to the climatic and weather conditions than qat, which introduces higher risks and more fluctuation in yield, quantity, quality and thus sales. Examples of such risks are warm winters, inadequate water application and improper irrigation scheduling. Moreover, the short harvest period of grapes results usually in over-supply and hence in lower market prices. Generally, it was observed that grape prices start higher at the beginning of the season and decline towards the middle of the season as supply increases. Nevertheless, prices are generally set by variety.

#### *6.3.3.2 Profits*

Profits are defined as the difference between total sales and total costs related to qat and grapes. Although qat is taxed, taxes are usually collected from sellers (middlemen) upon entrance to the city and in the marketplace, thus, this tax is carried by the consumers and not by the farmers. Figures 6.5 and 6.6 show the profits of qat and grapes from the pilot farm for the three seasons, respectively. Figures 6.5 and 6.6 show also the corresponding average water application levels.

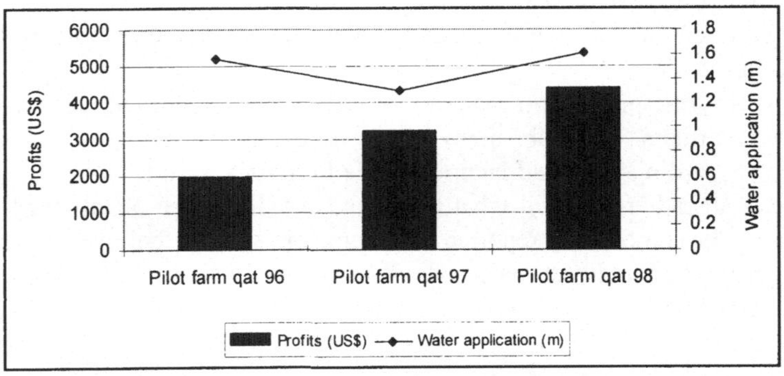


Figure 6.5 Profits from qat cultivation in the pilot farm.

Figures 6.5 and 6.6 show that profits from qat and grape cultivation increased in the period of 1996-1998 by 66% for both crops between '96 and '97 and 38% and 42% between '97 and '98 for qat and grapes, respectively. This rise in profits may be attributed to higher market prices of produce in general and cash crops in particular, which is the result of higher demand and an effort to offset the high rate of inflation. These results are also confirmed by the survey results: 75% of the surveyed farmers reported a continuous increase in profits in the period 1990-1996.

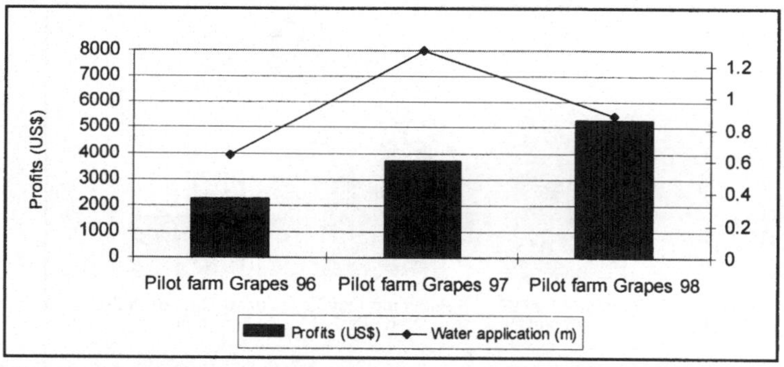


Figure 6.6 Profits from grapes cultivation in the pilot farm.

For qat, irrigation water application and profits were not correlated. Water application for qat in the three seasons was always higher than the irrigation water requirement, thus, it can be argued that the level of water application did not influence the yield and that the increase in profits is the result of higher market prices only. Although the qat harvest of the pilot farm was not measured[82], at an adequate level of water application it may be expected that the harvest quantity and quality per plot was comparable for the three seasons.

As for grapes, the increase in water application in the pilot farm in '97 contributed to higher (almost double) yield and considerable increase in profits. Although the grape yield in '96 on the pilot farm was much lower than the succeeding year, the unit-weight price of the harvest was around 30% higher. This may be attributed to lower supply; surveyed farmers reported a disease affected grape cultivation in some regions of the basin forcing many farmers to abandon the crop in the middle of the '96 season, so, the lower yield in the pilot farm in the '96 season is attributed to lower water application due to the old low yield pump and priority watering of the qat plots. The disease did not affect the pilot farm region. Although water application was reduced from 1.3 m in '97 to 0.8 m in '98, the resulting harvest was only marginally lower. Nevertheless, the unit-weight price of the harvest in '98 increased considerably by around 40% compared to that of the previous year, indicating a growing demand.

Figures 6.7 and 6.8 present profits as land and water profitability[83] aggregating all farm data.

Both land and water were increasingly profitable between 1996 and 1998, with a range of 0.8-1.3 $\$/m^2$ and 0.5-0.8 $\$/m^3$, respectively (Figure 6.7). The survey results (1996) show values similar to those of the '96 season on the pilot farm. Again, no correlation existed between water application and profitability, which is acceptable given that the water application during the three seasons exceeded the theoretical irrigation water requirements. It is noteworthy that land profitability exceeded water profitability in qat cultivation during all three seasons, which confirms sufficient water application.

[82] Qat harvest is difficult to measure by weight or volume.

[83] Land and water profitability are the total profits per unit of cultivated land and per unit of irrigation water application, respectively.

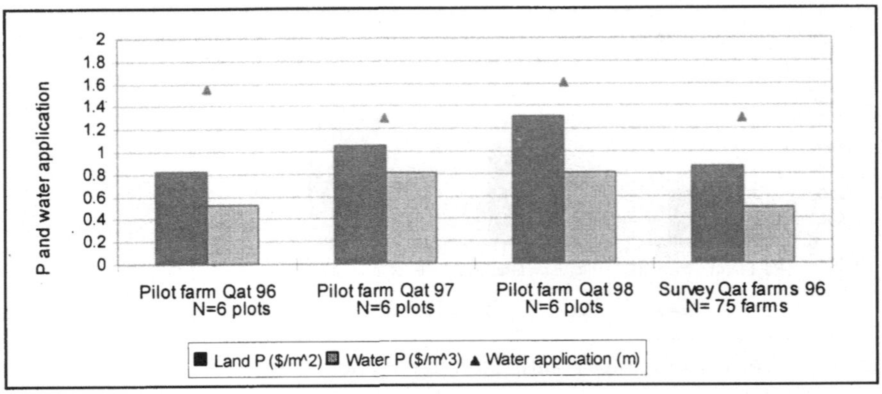


Figure 6.7 Land and water profitability (P) of qat cultivation in the Sana'a basin.

For grapes in the pilot farm, land and water profitability were 0.28-0.67 $/m^2$ and 0.36-0.76 $/m^3$, respectively (Figure 6.8). Unlike qat, water profitability for grapes exceeded land profitability in the '96 and '98 seasons suggesting low[84] water application. The switch in the '97 season was likely due to the near doubling of the water application resulting in higher yield. The increase in profits in 1997 however was comparatively small, hence lowering water profitability to below that of the previous year and lower than that of land. The survey results show a trend similar to those of the '96 and '98 seasons on the pilot farm.

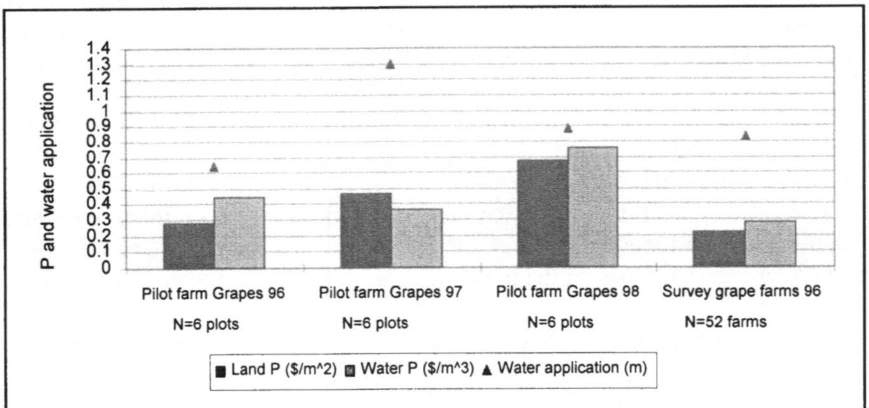


Figure 6.8 Land and water profitability (P) of grape cultivation in the Sana'a basin.

On the pilot farm land profitability for qat was twice that for grapes hence justifying the expansion of qat cultivation (between the vine rows and new three planting on two previously uncultivated plots). Basin wide, these results justify the trend in land use change, with most suitable land being converted and constructed for qat cultivation (Chapter 5).

[84] The water application for grapes as measured in the pilot farm and from the average reported for the surveyed farms was lower than the irrigation water requirement reported by HWC (1992). Doorenbos and Kassam (1979) report a total seasonal water requirement for grapes to vary between 500-1200 mm, depending on climate and the length of the growing season.

### 6.3.4 Cost-benefit analysis

The total cost of the inputs for water extraction constituted only 10-22% of the net profit from qat cultivation. The capital cost of water represented the larger share with 3.5-7.5% of the net profit (Figure 6.9).

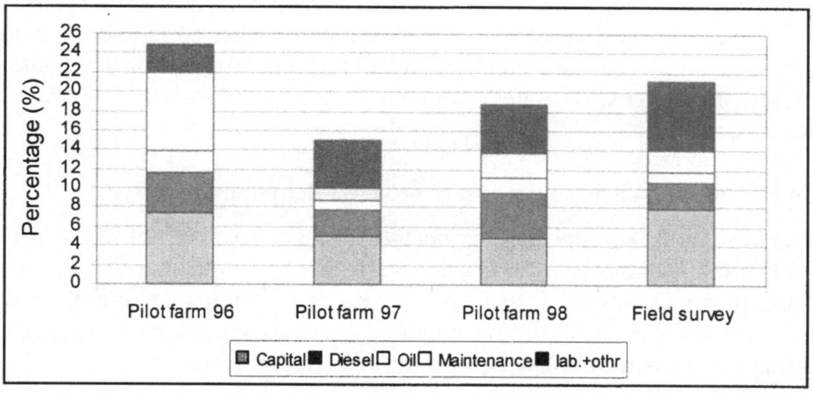


Figure 6.9 Ratios of the cost of the water related inputs to the profit from qat cultivation.

The cost of water amounted to 15-43% of the profit from grapes (Figure 6.10). Like for qat, the capital cost component for grapes had the larger share with a range of 5-14%. The higher proportion of the water cost in the profit from grape cultivation suggests that grape cultivation is more vulnerable to an increase in cost. In other words, if water costs are raised substantially, it is expected that farmers will react by lowering water application for grapes and may reduce the acreage under grapes. Although an increase in the cost of water and subsequent reduction in supply may trigger higher market prices for grapes, this rise may not necessarily result in proportionally higher unit profit given the nature of grape. Indeed, unlike grapes, qat is perceived by consumers to be a daily essential commodity and qat prices are likely to be less elastic than those of grapes. This is likely to further drive down the acreage under grapes.

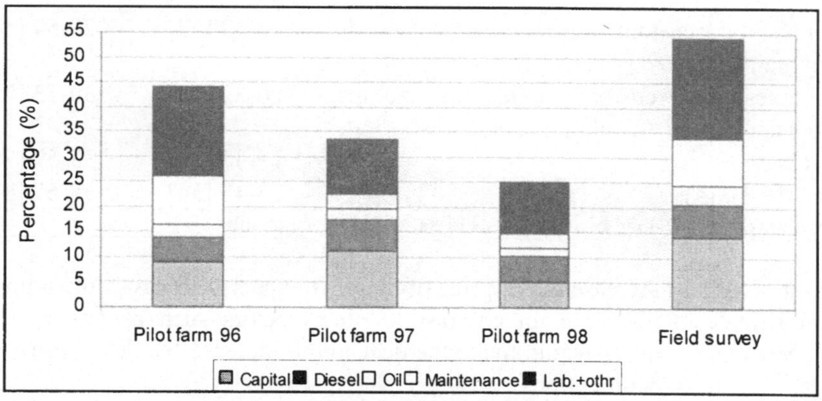


Figure 6.10 Ratio of the cost of the water related inputs to the profit from grape cultivation.

It emerges that the water related costs are relatively small when compared to the high profits from qat, and to a lesser extent grape, cultivation. For any increase in the pumping cost to be effective in water use reductions, it would have to be so large that it results in scaling down the profitability of cash crops. Higher water costs are expected to influence vegetable and grape cultivation before qat cultivation. In fact, it can be argued that higher input costs will speed up the current shift in cropping pattern to the water demanding qat, as farmers can pass higher costs to consumers. In addition, it is believed that qat farmers with their large profit margin can withstand large increases in the cost of water[85] without having to immediately recover their costs from the consumers.

### 6.3.5 Constraints to increasing the cost of groundwater pumping

The field data suggest that higher water costs will most probably not result in lower groundwater use. Nevertheless, the feasibility of alternatives to increase the cost of groundwater pumping deserves analysis: (1) lifting subsidies on diesel, and (2) increasing taxes and custom duties on imported groundwater abstraction equipment, and curtailing the spread of these equipment through import restrictions.

Assuming constant profit levels, the consequence of higher groundwater pumping cost according to different scenarios is shown in Table 6.3. Qat and to a lesser extent grapes are attractive crops with high profit potential (Figures 6.5-6.8). Even with higher input costs and constant profits, farmers have a relatively large profit margin that can withstand such increases. At constant profit levels, doubling the diesel price and maintenance costs would only marginally increase the cost-profit ratio. With further doubling of the capital cost, the cost-profit ratio increases to a maximum value of 45% and 67% for qat and grapes, respectively.

Table 6.3 Cost-profit ratio as a function of rising cost of groundwater pumping on the pilot farm (values between brackets pertain to the survey average).

| Action | Cost*-profit ratio (%) | |
|---|---|---|
| | **Qat** | **Grape** |
| No action | 15-25 [21] | 25-44 [54] |
| Lifting diesel subsidies (= doubling the price of diesel) | 18-29 [24] | 30-49 [60] |
| Higher taxes and custom duties on spare parts (= doubling maintenance cost) | 16-33 [23] | 28-54 [64] |
| Higher taxes and custom duties on pumps, generators and pipes (= doubling capital cost) | 20-32 [29] | 30-53 [68] |
| Actions (2)+(3) | 19-37 [26] | 33-59 [70] |
| Actions (2)+(3)+(4) | 24-45 [34] | 38-67 [84] |

* Includes capital, O&M, labor, and other costs of qat and grape cultivation

Significantly, a sharp increase in the price of diesel is expected to have only a limited effect on irrigated agriculture, in contrast to other sectors such as transport and industry. Such externalities would make the decision to increase the diesel price both socially and politically contentious.

[85] As one interviewed farmer answered the question "Will you reduce water application if diesel becomes more expensive?" by saying "I will not be stingy for a couple of thousand Reyals for a plot with a harvest worth of 60,000 Reyal".

The alternative of increasing the capital cost has fewer external effects. However, it was found that farmers have been turning to cheaper used and untaxed equipment in an effort to minimize costs (secondhand large pumps and old diesel engines stripped from old trucks). As described on chapter 5, the drilling rush took place mainly during the 1980s and it is expected that drilling intensity will further slow down in the coming years. With large reserves of used equipment in the market, increasing the cost of new pumping equipment is expected to yield only limited success in the short-term.

## 6.4 Conclusions and recommendations

Excluding the capital cost of land and initial plantation, the share of water related costs in the overall cultivation costs of qat and grapes was in the order of 60-70%. The average total annual cultivation costs were around 20 and 15 US¢/m$^2$ for qat and grapes, respectively. Of the water-related costs, the capital cost component constituted the highest share with 30-50%.

Diesel is the main energy source for groundwater pumping in the Sana'a basin with an average consumption of about 0.5 liter per m$^3$ of pumped groundwater. The annual financial burden on the government from diesel subsidy resulting from its use for groundwater abstraction in the Sana'a basin was estimated to be YR 0.9-1.4 billion (8.7-9.3 US$ million).

The 67% rise of diesel price in the end of 1997 resulted in an increase of the fuel cost component in the water-related costs from 28% in 1997 to 35% in 1998. The share of fuel (diesel) in the overall cultivation costs, the water related costs, and the O&M cost was 12-26%, 19-34%, and 29-56%, respectively.

Most farmers in the Sana'a basin were found to rely exclusively on crop sales for their income. Income range from crop sales was 0.8-2.2 and 0.3-1.3 US$/m$^2$ for qat and grapes in the pilot farm, respectively. The income range in the surveyed farms was larger with 0.1-4.1 and 0.03-2.3 US$/m$^2$ for qat and grapes, respectively, which may be explained by the wide range in harvest quality.

Profits from qat and grape cultivation have been rising since the early 1990s. Profits from qat were not affected by the fluctuation in the level of irrigation water application since water application for qat exceeded always the theoretical irrigation water requirement. Therefore, increasing profits reflect higher demand. As for grapes, the fluctuation in water application influenced yield and profits, especially since the water application for grapes has been mostly below the theoretical irrigation requirement level. Qat price on the market appeared un-elastic, in contrast to that of grapes. Sana'auites, the main buyers of the qat produces in the basin, see qat as a daily necessity without substitute. The qat market is a sellers' market. Qat prices, and thus profit, were influenced by seasonal supply: during the wet season the supply is temporarily satisfying demand and prices decline. Grapes, on the other hand, are produced during one season only, which tends to depress prices.

Land and water profitability was in the range of 0.8-1.3 US$/m$^2$ and 0.5-0.8 US$/m$^3$ for qat and 0.3-0.7 US$/m$^2$ and 0.4-0.8 US$/m$^3$ for grapes. The cost-profit ratio of qat and grape cultivation is in the range of 15-25% and 25-45%, respectively. The cost-profit ratio increased only from 15% in 1997 to 18.7% in 1998 for qat in the pilot farm despite the 67% rise in diesel price in 1997 and the increased water application by more than 20% in 1998. For grapes, the cost-profit ratio decreased from 34% in 1997 to 25% in 1998, due mainly to a substantial reduction in water application and 40% higher market price for the crop. The market creates clear incentives for farmers to irrigate qat by priority (often, over-irrigate) at the expense of grapes.

Increasing the input cost by doubling the diesel price would increase the cost-profit ratio by only 5%. Assuming constant profit levels, the probably unfeasible scenario of doubling the prices of fuel, maintenance, and capital components, would increase the cost-profit ratio by around 20%. This strongly suggests that measures to increase the input costs would not lead to lower water application on qat, but would probably further lower water application on grapes and vegetables.

The economic requirement that water should be used to maximize the return from its use in a sustainable manner conflicts with the actual dominance of water use for qat production. However, financial incentives targeted at the farmer's cost are likely to fail. To address the water scarcity in the Sana'a basin, priorities in water allocation should be made in a more participatory, explicit and transparent manner.

## 6.5 References

Beaumont P. (1985) irrigated agriculture and ground-water mining on the high plains of Texas, USA. *Environmental Concervation.* **12**(2), 119-130.

Carr T. (1977) Agricultural response to changing water prices in Arizona. M.Sc. thesis, The University of Arizona.

Doorenbos J. and Kassam A. (1979) *Yield response to water.* FAO Irrigation and Drainage paper 33, Rome, Italy.

Foppen J. (1996) Evaluation of the effects of groundwater use on groundwater availability in the Sana'a basin, Main report. Technical Report No.5, vol. 1, SAWAS-3, National Water and Sanitation Authority, Sana'a, Yemen.

Gardner R. and Young R. (1984) Effects of Electricity Rates and Rate Structures on Pump Irrigation: An Eastern Colorado Case Study. *Land Economics.* **40**(4), 352-359.

Heady E., madsen H., nicol K. and hargrove S. (1973) Agricultural water allocation, land use, and policy. *Journal of the Hydraulics Division, American Society of Civil Engineers.* **99**(HY10), 1795-1812.

Hexem R. and Heady E. (1978) *Water Production Functions for Irrigated Agriculture.* Iowa State University Press.

Howitt R., Watson W. and Adams R. (1980) A Reevaluation of Price Elasticities for Irrigation Water. *Water Resources Research.* **16**(4), 623-628.

HWC (1992a) Water Resources management Options in Sana'a Basin. Final Report Vol. 9, Technical Secretariat of the High Water Council, Sana'a, Yemen.

Kelly S. and Ayer H. (1982) Water Conservation Alternatives for California Agriculture: A Microeconomic Analysis. ERS Staff Report No. AGES820417.

Ministry of Legal Affairs (1973) Law No. (13) for the Establishment of a protection zone for the Sana'a Water supply. Official Gazette, No. 12.

NWSA (1996) Evaluation of the effects of groundwater use on groundwater availability in the Sana'a basin. Technical Report No.5, vol. 5, SAWAS-3, National Water and Sanitation Authority, Sana'a, Yemen.

Schiffler M. (1998) *The Economics of Groundwater Management in Arid Countries: Theory, International Experiencc and a case study of Jordan.* Frank Cass, London.

Shah T. (1997) Integrating Water Markets in Sustainable Water Resources Management. World Bank Research Observer.

Shumway C. (1973) Derived demand for irrigation water: The California aqueduct. *South, J. Agric. Econ.* **5**, 195-200.

Sloggett R. and Mapp H. (1984) An analysis of rising irrigation costs in the great plains. *Water Resources Bulletin*. **20**(2), 229-233.

# Chapter 7

## Multi Criteria Analysis on Options for Wastewater Reuse in Sana'a

### 7.1 Introduction

The rapid increase of the world population, particularly in the developing countries, is exerting a continuously growing pressure on the available water resources for municipal and industrial uses as well as for food production. Urbanization is yet another factor exerting increasing pressure on the quantity and quality of local water resources resulting from concentrated municipal and industrial water demand, and waste disposal. Between 1950 and 1990 the number of cities with population greater than one million jumped from 50 to 300 and it is projected to further rise to 400 by the end of the century. It is estimated that approximately 55 million people are added to the urban population of developing countries every year and that the number of cities with population between one and ten million inhabitants in Asia, Africa and Latin America has more than doubled since 1970 (UN, 1985; UN, 1997).

General water scarcity in arid and semi-arid regions is a major factor threatening the future sustainable development of many countries. As of 1995, 44 countries representing around 733 million people, of which more than half live in Africa and the Middle East, had below the stress level of 1,700 $m^3$/cap./a of renewable water supply (Postel, 1996). The alarming water shortage in many areas should be an incentive for greater, more careful water management that includes the still unconventional reuse of municipal wastewater.

Wastewater becomes a valuable resource when it can be renovated efficiently and economically to a suitably high quality, and where there is a market developing for such water. Arar (1991) lists the following specific reasons for the growing interest in wastewater reuse in developing countries:

1. High population growth, leading to greater quantities of wastewater.
2. Increasing environmental concerns reflected by stricter pollution control measures leading to higher quantities of wastewater to be treated at great expense.
3. Growing water scarcity in many arid and semi-arid regions of the world (increased demand and, hence, willingness to pay).
4. The nutrient value of wastewater adds attraction for agricultural use.
5. Depending on the degree of treatment, reclaimed wastewater is a reliably available resource that may be fit for irrigation, industrial, and municipal uses.

If not properly reused, wastewater should be considered an economic and financial liability since it deteriorates the environment by lowering the economic utility of high-value water resources, and as it requires expensive treatment for those resources to be restored to a quality suitable for further use. On the other hand, when wastewater is collected, treated and reused, in many cases it becomes an economic asset since it allows to replace and hence release higher quality fresh water for higher value uses.

The first six sections of this chapter review and discuss relevant literature on wastewater reuse, including the incentive system for wastewater reuse, the different types of reuse, treatment technology selection and financial and economic considerations. The second part of this chapter analyzes the reuse possibilities in Sana'a. This includes a review of the current local wastewater reuse practice, an assessment of the available wastewater in terms of both quantity and quality, identification of six reuse scenarios and five reuse criteria, and a multi-criteria analysis and valuation of the different scenarios.

## 7.2 Incentives for wastewater reuse

Incentives to reuse vary according to a complex set of national and local circumstances. Whereas in countries such as Japan and The Netherlands, pollution control regulations are the driving force for wastewater reclamation, in Israel, for example, water scarcity causes the relatively high percentage of wastewater reclamation, which is projected to contribute to around 36% of the national agricultural water supply by the year 2010 (Asano et al., 1996; Shelef and Azov, 1996). The following are four types of incentives for wastewater reuse.

### 7.2.1 Economic incentives

Economic incentives may be divided into macro and micro-level incentives. Macro-level incentives are those realized from the conservation of water resources or environmental quality by alleviating or delaying the need to develop expensive water supply augmentation projects. A second macro-level incentive is realized from the resulting reduction of environmental pollution, thus minimizing any future restoration costs. Micro-level incentives are those that trigger individuals to directly prefer, although sometimes at a risk, using renovated or raw wastewater. An example is the direct, unregulated use of raw sewage for agricultural irrigation.

### 7.2.2 Financial incentives

Even though wastewater collection and treatment are considered services for which beneficiaries are to be charged, in many cases, especially in the developing countries, full cost recovery is usually not realized. Nevertheless, in areas facing serious water scarcity, benefits from reuse may justify the additional treatment and distribution costs. Moreover, in other areas with strict effluent quality standards and restrictive discharge criteria, reuse becomes also financially feasible. Governmental policy can thus contribute to the financial feasibility of reuse projects. The Japanese government, for example, supports all reuse projects by subsidizing 50% of the total construction costs; in 1992, investment in wastewater treatment facilities constituted 0.7% of GNP and 8.6% of the total government investment in infrastructure (Asano et al., 1996). It may thus be concluded that many developing countries, with limited economic capacity to invest in expensive treatment facilities and with a low cost recovery potential, will find it increasingly difficult to justify reuse projects unless direct financial gains can be made.

### 7.2.3 Social incentives

With concerns growing over environmental protection and pollution control, the public approval of wastewater reclamation projects is generally rising. However, socio-cultural factors, e.g., perception of the health related risks to bodily contact, and the educational and awareness levels of the public, strongly influence the extent and success of wastewater reuse projects. Some of these perceptions and habits may have been codified in religious and traditional teachings.

### 7.2.4 Technical incentives

Thanks to rapid advancement, partly the result of the adoption of stricter environmental policies, treatment technology is no longer considered the limiting factor in wastewater reuse projects. Nevertheless, appropriate technology selection to suit the local conditions increases the sustainability and financial feasibility of reuse projects. In the developing countries, with limited financial capacity and abundant unskilled labor, low-technology solutions are usually preferred, which commonly means less mechanization, lower degree of automation and maximum involvement of the local community in the different phases of project implementation (planning, construction, operation and maintenance) (Veenstra et al., 1997).

## 7.3 Types of wastewater reuse

### 7.3.1 Selection criteria

Treated wastewater is used in the three main water using sectors: agriculture, municipal and industry. Five general characteristics of wastewater influence decisions concerning reuse:

- Quantity of available wastewater: This is influenced by population, the per capita water consumption, and the sewerage coverage.
- Quality of wastewater: Effluent quality depends greatly on the tap water composition from which the wastewater derives, the type of wastes carried, and the type and degree of treatment.
- Availability and costs of alternative water sources: The easily accessible and cheaper water resources will be developed first, leaving the more expensive alternatives of, for example, import or desalination to the last. The cost of reuse alternatives typically falls between the two extremes.
- Availability of funds: This depends to a large extent upon governmental policies toward infrastructure subsidies and the degree of local cost recovery.
- Public attitude towards wastewater reuse.

### 7.3.2 Wastewater reuse for agriculture

Wastewater reuse for agricultural irrigation is an old and relatively widespread practice not only in the water short regions, but also in water rich countries. In these cases, the economic rationale is based on avoiding the cost of the deterioration of the rivers that would otherwise be polluted. In Paris, wastewater irrigation dates back to the nineteenth century, and currently still more than 2,000 hectares are irrigated with treated wastewater (Bontoux and Courtois, 1996). In 1966, the city of Tallahassee, Florida, started a wastewater reuse program for agriculture; in 1992, 81,400 $m^3/d$ of secondary-quality effluent irrigated approximately 700 hectares of corn, soybeans, coastal grass and rye (USEPA, 1992). Table 7.1 shows some examples of wastewater reuse in agriculture.

Table 7.1 Examples of wastewater reuse in agriculture (after Bartone and Arlosoroff, 1987).

| Location | Volume reused ($Mm^3/a$) | Cultivated area | Comments |
|---|---|---|---|
| Germany | 100 | $\geq 28*10^3$ hectares | 3% of the total sewage production. |
| China | 10,000 | $1.33*10^6$ hectares | 27% of the total sewage production. |
| Mexico City | 1,500 | $90*10^3$ hectares | 100% of the total sewage production. |
| Santiago, Chili | 190 | $16*10^3$ hectares | 100% of the total sewage production. |

Advantages of raw/treated wastewater reuse for irrigation can be summarized as follows:

- An increase in agricultural production, as in many cases, acreage under rain-fed farming is converted to wastewater irrigation.
- Utilization of the nutrient content of wastewater, which results in a reduction of fertilizer application; the economic value of the nutrients in the reused wastewater has been estimated to range 3-5 US¢/$m^3$ (Shuval et al., 1986).

*7.3.2.1 Standards and guidelines of wastewater reuse in agriculture*

Even though environmental protection and water supply augmentation are the primary objectives of many reuse projects, regulations, standards and guidelines usually reflect health concerns. Public health protection is usually achieved by (1) reduction of pathogenic organisms, (2) control of the chemical constituents of wastewater, and (3) limitation of human exposure to wastewater.

The connection between disease outbreaks and wastewater reuse for irrigation was realized early this century. As early as 1918 regulations in California, USA, prohibited the use of raw sewage for irrigation, and by 1948 through the Water Pollution Act, a reuse permit system for wastewater reuse was introduced. Public health concerns led to high quality standards for wastewater reuse. The California standards call for the total elimination of pathogens in wastewater regardless of the health hazards associated with partial removal during treatment. The 1973 WHO guidelines for wastewater reuse adopted a similar conservative approach requiring nearly complete coliform removal for irrigation, regardless of the crop type.

These guidelines recommended high removal efficiencies and consequently high treatment costs that could not be achieved by most developing countries, and, hence, they hindered wider application of wastewater reuse in agricultural irrigation under proper regulations. Eventually, the WHO took a more realistic approach, in which health risks were evaluated against epidemiological evidence, taking into account the technical and economic capacity of the developing countries. After reevaluation, it was concluded that the main risks associated with wastewater irrigation are those connected to helminthic diseases (Shuval et al., 1986). These findings resulted in the adoption of the much milder WHO guidelines for wastewater irrigation in 1989 (Mara and Cairncross, 1989). An important justification for these guidelines was the fact that most conventional irrigation water sources contain higher coliform counts than those required by the reuse guidelines, as 50% of the world's rivers contain fecal coliforms counts ≥1000 per 100 ml (Shuval, 1991).

Even though the WHO guidelines are easily achievable using minimum treatment, many countries are still adopting more stringent standards for wastewater irrigation as can be seen in Table 7.2. Criticism of the current WHO guidelines that they are too lenient and provide insufficient safeguards to public health, come mainly from arguments that the origin and composition of indicator organisms in natural rivers totally differ from those found in municipal wastewater (Shelef, 1991; Shelef and Azov, 1996). In addition, even though wastewater irrigation brings extra benefits to agriculture in the form of nutrient supply, the effects of increased salinity and toxicity on both soil and crops should be evaluated on a case by case basis.

Table 7.2 Representative standards for wastewater reuse in agricultural irrigation (Mara and Cairncross, 1989; USEPA, 1992) ("Restricted irrigation" refers to irrigation of trees, industrial crops, fodder crops, fruit trees, and pasture. "Unrestricted irrigation" refers to irrigation of edible crops, sports fields, and public parks)

| Country | Restricted irrigation | Unrestricted irrigation |
|---|---|---|
| Oman | Maximum 23 TC/100ml<br>Average 2.2 TC/100ml<br>Greenbelt irrigation: <10,000 TC/100ml. | Crop irrigation not permitted. |
| Kuwait | <10,000 TC/100ml | <100 TC/100ml, but not for salad crops or strawberries. |
| Saudi Arabia | Use of secondary effluent permitted for forage crops, field crops and vegetables that are processed and also for landscape irrigation. | <2.2 TC/100ml<br><50 FC/100 ml |
| Arizona, USA | Reuse for non-food crops:<br>100 FC/100ml (median)<br>4000FC/100ml (single sample) | Reuse for food crops:<br>2.2 FC/100ml (median)<br>25 FC/100ml (single sample)<br>Reuse for processed food:<br>1000 FC/100ml (median)<br>2500 FC/100ml (single sample) |
| California, USA | Fodder, fiber and seed:<br>primary treatment required<br>Pasture for milking animals:<br>23TC/100ml (median) | 2.2 TC/100ml (median)<br>23 FC/100ml (single sample) |

TC Total coliform
FC Fecal coliform

### 7.3.3 Wastewater reuse for municipal uses

Wastewater reuse for municipal purposes can be divided into two main categories: potable and non-potable uses. Potable reuse of wastewater is further divided into direct and indirect uses. Direct reuse implies that wastewater is treated to a level acceptable for human consumption. Reclaimed wastewater for direct potable use usually goes through two separate conventional and advanced treatment processes.

Indirect potable reuse of wastewater is more common and in fact often unwittingly applied through the disposal of previously treated wastewater into surface or groundwater, which is used downstream or down gradient as source for potable water supply.

#### *7.3.3.1 Direct potable reuse*

The city of Windhoek, Namibia, is one of the very few examples of direct reuse, and has had a wastewater reuse program for direct potable use for 25 years. The growing local rural agriculture water demand around Windhoek has exerted increasing pressure on the available fresh water resources for municipal water supply, and thus the option was recently taken to expand the previous 4,800 $m^3$/d reclamation capacity to the maximum of 21,000 $m^3$/d. An extensive monitoring program indicated no health hazards associated with the use of reclaimed water for potable purposes in Windhoek. Moreover, continuous monitoring and research have resulted in four modifications and improvements in the advanced treatment system since 1968 (USEPA, 1992; Haarhoff and Van der Merwe, 1996).

In a 44 l/s demonstration/pilot plant in Denver, USA, research has been conducted for more than ten years in the area of wastewater reclamation for direct potable use. The system treats disinfected effluent from a tertiary treatment plant using advanced treatment processes that include lime induced clarification, re-carbonation, filtration, ultraviolet irradiation, activated carbon adsorption, and reverse osmosis, in addition to ozonation and chlorination. The product water is of a high quality that meets all drinking water standards. Given the water's origin, regulation requires that the treatment processes be specified in every case if the reclaimed water is to be used for potable uses in order to avoid public health hazards (Bouwer, 1992).

Even though the experiences of Windhoek and Denver have proven the viability of the technology to produce acceptable drinking quality water from wastewater, it is unlikely that it will be widely adapted for the following two main reasons:

- Costs: It is very expensive to treat wastewater to drinking water quality, especially if compared with transportation and treatment costs of other more conventional water resources. Generally speaking, it is more cost-effective to reuse partially treated wastewater for industrial or agricultural purposes, thus "releasing" high-quality fresh water for potable use.
- Public acceptance: While most people accept and may encourage non-potable, or even indirect potable uses of reclaimed wastewater, the vast majority of the public tends to be reluctant to accept wastewater reuse for direct potable use.

#### *7.3.3.2 Indirect potable reuse*

Many of the large cities and towns that lie along the major rivers and lakes depend on water from those rivers for their domestic water supply. These water bodies at the same time receive treated and raw sewage from upstream cities and industries. The lower reaches of the river Rhine in Germany, which serve as a raw water source for around 600,000 people, often contains 40% of (treated) sewage effluent, which may rise to 100% during periods of very low flow. At average flow, the river Thames in the UK consists of 14% of treated wastewater. During times of drought, the water supply source for the city of Agra, India, and that for other cities along the Yamuna river, consists entirely of partially treated sewage from the city of New Delhi (Ali, 1987). Many urban centers in developing countries (e.g., Jakarta, Gaza, etc.) rely on local ground infiltration of domestic wastewater; unwittingly, a substantial part of the water is pumped up again with the groundwater for household purposes (chapter 2; Abuzahrah, 1995). Raw sewage from around 80% of the city of Sana'a, Yemen, is directly discharged to the under ground, which to a certain extent has changed the quality of the local groundwater aquifer, the prime source for the city's water supply (Al-Hamdi, 1997; Chapter 3).

Groundwater recharge with treated wastewater is used in some places in the world, of which the recharge projects in Los Angeles, California, Orange County, California, and El Paso, Texas, are among the largest. The recharged aquifers are usually used as water supply sources. In Los Angeles, in 1990 a total of 62 $Mm^3$ of treated wastewater, which accounts for 30% of the total recharge flow, was used to recharge the local aquifer using more than 260 hectares of spreading basins. Water Factory 21 in Orange County injects around 56,000 $m^3$/d of advanced treated wastewater to recharge a coastal aquifer, thus augmenting water for further municipal use, and at the same time creating a fresh water barrier against saline water intrusion (Asano, 1985; USEPA, 1992). After conventional secondary wastewater treatment, 27,000 $m^3$/d of treated wastewater effluent from the city of El Paso, Texas, is diverted to an advanced treatment plant. The final effluent is injected to recharge the local aquifer, where it is pumped for municipal supply from production wells 3 km down-gradient of the injection wells (Bouwer, 1992).

The dilution in the "natural" water body, and the self-purification of those water bodies, seem to favorably change public attitude towards indirect potable reuse. Groundwater is generally perceived to be of a good quality which may give rise to the prejudice that groundwater is protected against the consequences of wastewater discharge to the aquifer. In the industrialized countries, regulations for groundwater recharge with reclaimed wastewater are much stricter than those required for discharge of reclaimed wastewater to surface water. At present, there is a growing interest to utilize the modest self-purification of water flow in the soil layers, through the use of the Soil Aquifer Treatment (SAT) (USA and Israel), river bank infiltration (Germany and The Netherlands), and dune infiltration (The Netherlands) (USEPA, 1992). The required quality of reclaimed wastewater prior to recharge needs usually to be evaluated on a case by case basis.

#### *7.3.3.3 Non-potable reuse*

Non-potable reuse of reclaimed wastewater is one strategy that holds possibly more promise, since it (1) tends to require less expensive treatment, (2) is associated with lower health risks, (3) faces minimal public rejection, and (4) still fully supports resource conservation. Non-potable municipal reuse of reclaimed water can be divided into two general categories, namely outdoor uses and indoor uses. Outdoor uses include the irrigation of landscape, greenbelts, golf courses, public parks, sport fields, etc., in addition to fire-fighting, while indoor uses include toilet flushing, train washing, etc.

The use of treated wastewater for irrigation of public parks, sport fields and recreational sites has become recently a relatively widespread practice. In 1987, 60,000 $m^3$/d of treated sewage was used to irrigate shelter belts, forestry lots and greens along roads in Kuwait (Arar, 1991). Treated wastewater from the city of Abu Dhabi, United Arab Emirates, has been used for irrigation of municipal areas since

1976, where a low pressure distribution system avoids the health risk associated with high pressure sprinkler systems. In 1991, 920 l/s of treated wastewater from the city of St. Petersburg, Florida, provided irrigation water through a dual distribution system to irrigate individual home gardens, parks, school gardens and golf courses (USEPA, 1992). The seaside Chinese cities of Dalian, Tienjin, Hong Kong and Shenzhen are experimenting with reuse in hotels and other large-scale water using establishments.

Wastewater reuse gains importance in arid regions because of the obvious scarcity value of water. But also in non-arid regions, the demand for and, hence, the opportunity value of high quality freshwater are sufficiently elevated to pay for treatment and reuse. Japan, with an average precipitation of 1,700 mm/a, has launched comprehensive urban wastewater reuse projects since 1968. In the reclamation project of the Shinjuku district, in 1994 an average of 2,700 $m^3$/d of reclaimed wastewater was delivered to 19 high rise buildings for toilet flushing through the use of a dual distribution network. In the city of Fukuoka, an average of 3,300 $m^3$/d of treated wastewater was used to supply 215 installations with toilet flushing water during 1994. During 1993, it was estimated that 111,000 $m^3$ of reclaimed wastewater from the Shibaura wastewater treatment plant in Japan was used for train washing (Asano et al., 1996; Maeda et al., 1996).

### 7.3.4 Wastewater reuse for industrial and environmental uses

The availability of well treated wastewater at comparatively low cost, and the scarcity of water resources in general or of the good-quality water, are strong incentives for innovative reclamation projects. Since 1951 effluent from the Mikawashima treatment plant in Tokyo, Japan, has been used as a water source for a local paper mill. This reuse came about as a result of unacceptable quality of the local natural surface water source (Asano et al., 1996). In Saudi Arabia, the Riyadh wastewater treatment plant produces over 250,000 $m^3$/d of effluent, which is among others used as cooling water, process water for crude oil desalting, and for boiler feed (Arar, 1991). In Baltimore, Maryland, 4,380 l/s of treated wastewater from the city is used as cooling water for the Bethlehem Steel Company. Thirty five percent of water demand for power generation in the state of Nevada, USA, (3,940 l/s) is met by secondary wastewater effluent from the city of Las Vegas (USEPA, 1992). During drought conditions, industries in Chennai, India, are served with low priority service resulting in frequent interruptions. Two major industries in the city, Madras Refineries Limited and Madras Fertilizers Limited have taken the city's proposition to buy uninterrupted secondary treated sewage flow for the price of Rs. 4/$m^3$ (US4 11.5/$m^3$), and invest in advanced treatment to upgrade it to their desired higher quality. Consequently, reliance on groundwater for water supply of these industries has dropped by 55% between 1992 and 1996. Although the total financial cost of renovated water is about 60% higher than that supplied by the Madras Metropolitan Water Supply and Sewerage Board (MMWSSB), the additional economic benefits favored the option of wastewater renovation. After successful supervised experiments, MMWSSB also decided to develop 100 ha of sewage-irrigated agro-forestry, where it was established that the timber value in five years would generate twice the income of the previously irrigated cattle fodder paragrass. Plans were being prepared to expand this project to reach 400-500 ha using around 100,000 $m^3$/d of treated sewage (Nair and Selvam, 1996). Reuse of wastewater for industrial purposes may require high degrees of treatment, thus incurring high costs. Nevertheless, water scarcity and environmental policies may increase the economic feasibility for such reuse.

The use of treated wastewater to reduce, halt, or even reverse saline water intrusion in coastal aquifers is an example of environmental protection. The problem of groundwater degradation resulting from seawater intrusion in Orange County, California, was solved by implementing a dual-purpose artificial recharge project using reclaimed wastewater (Asano, 1985). A 32 US$ million reclamation project in Salalah, Oman, consisting of partial wastewater collection, an advanced wastewater treatment plant, and a groundwater recharge system, is expected to produce 20,000 $m^3$ of daily effluent. It was concluded that the project will decrease pollution disposal to the environment, reduce the current seawater intrusion, and will avoid the need for development of costly water supply options like desalination (Walsh, 1996). In the clean stream restoration project in Japan, reclaimed wastewater was supplied to restore the Nobidome irrigation channel. In total, 12.4 $Mm^3$ of treated wastewater from the Tomegawa-Joryu treatment plant was devoted to stream augmentation during 1993 (Maeda et al., 1996).

## 7.4 Treatment technology selection

The treatment to be adopted for reclamation depends mainly on the wastewater composition, the standards to be met and the particular intended purpose of the effluent. Complexity of technology varies widely between the simple, low cost oxidation ponds to the expensive reverse osmosis (Figure 7.1). In most industrialized countries, where discharge standards are strict, secondary treatment (using activated sludge or trickling filter plants) is usually applied followed by tertiary filtration and disinfiction for reuse in agriculture, or for non-potable municipal or industrial purposes. For potable uses of treated wastewater or for groundwater recharge through direct injection, advanced treatment is usually necessary. In most cases the advanced treatment plant is completely separated from the conventional preceding secondary treatment plant. The advanced treatment plant in the city of Windhoek, Namibia, includes chemical clarification, dissolved air flotation, chlorination, sedimentation, sand filtration, break-point chlorination and carbon filtration. The resulting water is blended with water from freshwater sources and delivered as drinking water to consumers (Haarhoff and van der Merwe, 1996). The product water from the Water Factory 21 in the Los Angeles County wastewater reclamation project, is blended with fresh groundwater and used for groundwater recharge through deep well injection (Asano, 1985).

On the other side, many developing countries apply raw sewage reuse, mostly in agriculture. This comparatively widespread practice is completely demand-driven and commonly benefits poorer farmers. In these cases the financial incentives for the farmers outweigh both the prohibitive governmental regulations and the presumed traditional or social bias against wastewater reuse. This is found in China, Yemen, Syria, Morocco, Tunisia, Kenya, India, Chile, Mexico, etc., where raw sewage or partially treated effluent is used locally for fodder grain or even vegetable cultivation (UN, 1985). For developing countries, the 1989 WHO guidelines for wastewater reuse in irrigation assist to develop regulated and organized reuse projects. In areas where land is available at reasonable cost, oxidation ponds are usually preferred as treatment prior to reuse due to their low operation and maintenance requirements and high reliability. Nevertheless, in areas where land is expensive, it may be a costly option (Alaerts et al., 1991). Shuval et al. (1986) suggested that the use of a first stage 1-2 days anaerobic pond followed by a 7-10 days facultative ponds is adequate to remove virtually all helminths and reduce the fecal coliforms to below 1000 colonies per 100 ml (WHO guideline for unrestricted irrigation). Duckweed-based ponds and fish ponds provide a tertiary-quality effluent while at the same time they create the possibility to become financially self-sustaining by providing valuable protein-rich biomass for animal feed and fish for human consumption (Skillicorn et al., 1993; Alaerts et al., 1996).

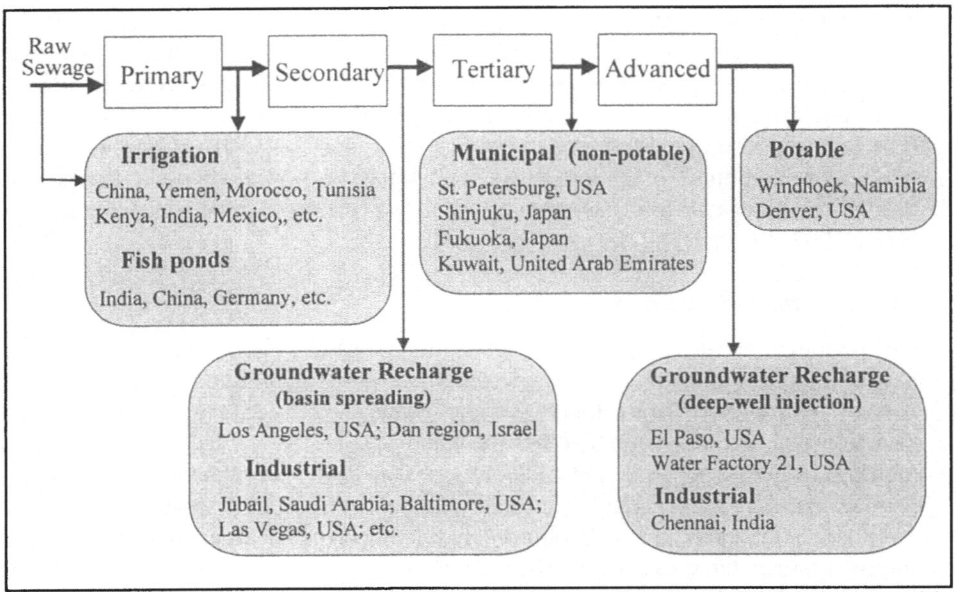


Figure 7.1 Schematic presentation of reuse options and treatment technologies, with examples ("advanced treatment" covers tertiary and in some cases includes reverse osmosis).

The use of SAT is gaining interest as a polishing process prior to reuse. SAT makes use of the natural ability of the soil particles to purify wastewater during its passage through the sub-soil medium. Quality improvements result from, notably, screening, filtration, sedimentation, adsorption and microbiological metabolization. The SAT system consists of infiltration basins, where the treated effluent is allowed to infiltrate and reach the local groundwater, to be recovered by pumping down-gradient of the infiltration site. Quality improvement is evident from the SAT in Tel Aviv (Dan region), Israel, where significant reductions in BOD (from 29 to <0.5 mg/l) and fecal coliforms (from $4*10^5$ to <1.1 per 100 ml) are achieved (Shelef, 1991). However, for higher efficiency, SAT requires low turbidity feed-water to minimize clogging. When treated wastewater is used for infiltration, infiltration rate depends on the effluent quality, soil type, geology, hydrogeology and climatic conditions. Using secondary effluent, Bouwer (1992) argues that the use of SAT followed by reverse osmosis would produce water fit for potable purposes, thus saving over 50% of the costs of the alternative advanced treatment. Moreover, SAT is considered a "robust and failure free" system that does not require highly skilled labor for operation. He also suggests that due to the high capacity of SAT to reduce BOD, prior secondary treatment might not be necessary. It was also suggested that the passage of treated wastewater through the underground lowers the psychological rejection increasing the public's support for reclamation projects.

## 7.5 Socio-cultural perception

Public opinion and acceptance of wastewater reclamation are considered critical factors in wastewater reclamation projects. Many studies into the subject of public opinion and acceptance depend on contingency valuation surveys. This survey technique is widely applied for many purposes through it is viewed by some as unreliable since in many cases the respondent's reply to hypothetical questions could differ from his actual reaction when faced with the real situation. Nevertheless, those studies have identified some factors to play an important role in the decision to accept wastewater reuse (Table 7.3).

Bruvold (1988) suggests that the studies (mentioned in Table 7.3) are incomplete since they propose generalized reuse options for undefined time and place frames. In a survey that took into account the proposed type of reuse for the respondents' particular communities in a specified time frame (a realistic or ongoing project), he concluded that the results in general didn't support conclusions of previous studies. He identified five factors that play an important role: (1) health effects, (2) environmental effects, (3) treatment costs, (4) distribution costs, and (5) water conservation. It was however, recommended that more research should be carried out to develop reliable general guidelines for wastewater reuse projects.

The view of Islamic tradition towards water seems to be of a ritual[86] nature, while the concerns with health aspects are left to be resolved through the growing human knowledge. In Islam, water has been characterized into three main categories, namely *tahur*, *taher*, and *mutanajjs*. *Tahur* is fit for ritual uses and considered the cleanest of the three. Although not fit for ritual purposes, *taher* is considered to be clean enough to be used for mundane uses such as cooking, washing, bathing, etc. The third category, *mutanajjs*, is considered unclean and not fit for use due to contamination that has changed one or more of its properties (color, taste or odor). Nevertheless, both *taher* and *mutanajjs* water can be converted into *tahur* water if adequate dilution with *tahur* water takes place, as the resulting water has the general properties of the *tahur* water. Removal of impurities through treatment can also convert *taher* and *mutanajjs* water into *tahur*. A body of Muslim scholars, the Organization of the Eminent Scholars of Saudi Arabia, has approved the reuse of wastewater after adequate treatment for all purposes including ritual uses (Wilkinson, 1978; Farooq and Ansari, 1983). The increasingly positive attitude of Islam towards wastewater reuse may be viewed as an encouraging factor for public acceptance.

---

[86] Mainly water use for ablution.

Table 7.3 Factors affecting acceptability of wastewater reuse (all in USA communities, '70s-'80s).

| Factor | Details | References |
|---|---|---|
| (1) Degree of bodily contact | The degree of acceptance to reuse reclaimed wastewater is inversely proportional to the intensity of bodily contact. While only 50% of the surveyed people indicated an acceptance to drink or cook with treated wastewater, around 95% accepted the idea of reuse for industrial cooling and golf course irrigation. | Baumann and Kasperson, 1974; Sims and Baumann, 1974; Bruvold, 1988. |
| (2) Educational and awareness level | Higher education increases the confidence in science and technology, thus increasing the acceptance level for wastewater reuse. A minimum knowledge or awareness about wastewater treatment and reclamation has a significant positive influence on the respondent's attitude. | Baumann and Kasperson, 1974; Sims and Baumann, 1974; Bruvold, 1988. |
| (3) Age and income | Age was correlated with the acceptance level. Acceptance of respondents older than 50 years to use treated wastewater for drinking was 25-30% compared to 50% acceptance level for those younger than 50 years. Higher income respondents indicated higher acceptance for wastewater reuse (this cross-correlates with item 2). | Baumann and Kasperson, 1974; Sims and Baumann, 1974. |
| (4) Opinion of reference group | Opinions of recognized experts (medical doctors, health officials, engineers and scientists) were found to greatly influence the public opinion regarding wastewater reuse. However, divergent views concerning health hazards, water quality, or economics of wastewater reuse among the local experts would lead to a "confused public", resulting in total rejection. | Baumann and Kasperson, 1974; Sims and Baumann, 1974 |

## 7.6 Economic and financial considerations

Costs limit the investment in reuse technology in areas where fresh water resources are readily available and thus cheap. As water resources become scarcer, especially in the urban areas of arid and semi-arid zones where the alternative of water import from other regions is associated with expensive conveyance infrastructure and rising competition and conflicts, wastewater reclamation becomes a cost-effective alternative. Similarly, stringent effluent quality standards also promote wastewater reuse as they raise the discharge costs and narrow the cost differential between treatment for discharge, and reuse.

Municipal wastewater reuse projects often are high in capital cost, since besides the expensive sewerage system and treatment facilities, storage and a conveyance and distribution network need to be included. Table 7.4 shows storage and distribution facilities in four wastewater reuse projects in the United Arab Emirates (UAE), Qatar, and Saudi Arabia. Bartone (1994) reports that the post-treatment for the wastewater reclamation project in Irvine Ranch Water District, California, USA, constitutes only 24% of the total investment, while 43% of the cost is attributed to the distribution network. For the reuse project in Dubai (UAE) the capital cost for treatment (incl. tertiary treatment) is around 27% of the total investment, while the share of the delivery and distribution network is in the order of 50%.

Table 7.4 Storage and distribution of wastewater reclamation projects for agriculture in the Middle East (Banks, 1991).

| | **Abu Dhabi UAE** | **Dubai UAE** | **Taif Saudi Arabia** | **Doha Qatar** |
|---|---|---|---|---|
| Quantity | 115,000 $m^3$/d | 26,000 $m^3$/d | 20,000-67,000 $m^3$/d | 67,000 $m^3$/d |
| Storage | 83,000 $m^3$ | Elevated 4,000 $m^3$<br>Ground 40,000 $m^3$ | Elevated 20,000 $m^3$<br>Ground 100,000 $m^3$ | Elevated 9,000 $m^3$<br>Ground 10,000 $m^3$ |
| Distribution | 110 km of pipes, of which 50 km are 200 and 600 mm diameter mains, 20-25 tankers, many thousands of sprinklers. | 200 km mains, 1,000 km laterals, 10,000 sprinklers, 80,000 sprays, 3000 km of drip irrigation lines, and 14 tankers. | Transmission lines, drip irrigation lines, and side roll sprinklers. | Transmission lines, laterals, and 15 center pivot sprinklers serving 42.5 hectares each. |
| Charge for effluent supplies | Free | Free | Free | Free |

While the oil rich Gulf states can afford to construct the necessary infrastructure (Table 7.4), the product water is delivered to the farmers free of charge with full governmental subsidy (Banks, 1991). Within these countries the incentive for such projects may include pollution control, water resource conservation and an encouragement policy towards agricultural production. Given the high costs and the low economic return from the reuse of reclaimed wastewater in agriculture, it is doubtful whether this type of reuse is the most cost-effective, as reuse for industrial purposes or groundwater recharge for municipal uses may be from an economic view point more competitive.

Whereas the reuse projects in the Gulf states do not aim at cost recovery, projects elsewhere typically incorporate partial or full cost recovery. In 1994, the unit price of water from the reclamation project in the Shinjuku district of Tokyo, Japan, (reuse for toilet flushing) was around ¥540/$m^3$ (5.4 US$/$m^3$), compared with ¥667/$m^3$ (6.67 US$/$m^3$) for potable municipal supply. Likewise, the wastewater reclamation project for toilet flushing in the city of Fukuoka, Japan, was priced at ¥350/$m^3$ (3.5 US$/$m^3$), 16% lower than the domestic water supply. Still, connection fees and user charges for reclaimed wastewater in Japan recover only 50% of the capital costs plus all recurrent costs (Maeda et al., 1996; Asano et al., 1996).

Table 7.5 summarizes the unit costs of reclaimed wastewater at different locations compared to the desalination costs as the most expensive, extreme alternative. A breakdown of associated costs of wastewater treatment processes in the USA is shown in Table 7.6.

Table 7.5 Total costs of wastewater treatment vs. desalinization cost of local salty or brackish water in different locations. (UN, 1985; USEPA, 1992; Al-A'ama and Nakhla, 1995).

| Location | Unit cost of treatment (US$/m$^3$) | Unit cost of desalination (US$/m$^3$) |
|---|---|---|
| United Arab Emirates (1982) | 0.3 – 0.41 | 1.0 – 1.45 |
| Qatar (1981) | 0.24 | 1.45 – 1.64 |
| Bahrain (1984) | 0.84 (incl. RO) | -- |
| Bahrain (1984) | 0.28 (excl. RO) | -- |
| Al-Jubail, Saudi Arabia (1995) | 2.2 (up to tertiary treatment) | 2.67 |
| Windhoek, Namibia (1980) | 0.31 (advanced treatment only) | -- |
| Denver, USA (1988) | 0.6 (advanced treatment only) | -- |
| USA (1980) | 0.13 – 0.32 | 0.87 |

RO: reverse osmosis

Table 7.6 Overall (capital + recurrent) average treatment costs in the USA (UN, 1985)

| Type of Treatment | Unit cost (US$/m$^3$) |
|---|---|
| Secondary treatment | 0.16 |
| Secondary treatment + filtration | 0.19 |
| Secondary treatment + filtration + activated carbon (advanced treatment) | 0.25 |
| Advanced treatment + reverse osmosis | 0.75 |

The unit costs are based on a 37,850 m$^3$/d capacity plant and according to 1980 price levels.

Since most wastewater reuse projects are directed towards the agricultural sector, sustainability of those projects should take into account the ability and willingness of farmers to contribute to the costs of treatment and distribution. Experience in Peru indicates that local farmers may be willing to compensate for some O&M tasks associated with treatment, storage, and the delivery system through in-kind contributions (Bartone and Arlosoroff, 1987). Nevertheless, it seems that contributions by farmers to recover part of the costs of reclamation projects may only be successful if farmers accept the principle of cost recovery and if other water resources are scarce and expensive.

## 7.7 Purpose, approach and methodology

### 7.7.1 Purpose

The remaining sections of this chapter analyze the wastewater reuse options in Sana'a. This includes a review of the current local wastewater reuse practice, an assessment of the available wastewater in terms of both quantity and quality, identification of six reuse scenarios and five reuse criteria, and finally a multi-criteria analysis and valuation of the different scenarios.

### 7.7.2 Approach and methodology

The limited water use in Sana'a in the past, was a direct result of water shortage and resulted in insignificant quantities of wastewater that posed no threat to either the human health nor to the local environment. As water became more accessible since the early 1970s, domestic wastewater disposal had an increasing damaging impact. In order to analyze the current practice of wastewater reuse in the Sana'a area, it is important to clearly identify the different disposal methods of domestic wastewater. At present, around 20% of the city's population is connected to the sewerage system, which discharges into oxidation ponds for partial treatment, while the remaining 80% of the city's population still depends on cesspits for direct sewage disposal.

Effluent of the ponds (and bypassed raw sewage) flow along a natural channel where it is mainly used for irrigation of mostly fodder and cereal crops. Sewage from cesspits infiltrates through the subsoil layer to reach the groundwater aquifers. Even though the extensive use of cesspits has resulted in groundwater contamination under the heavily populated areas of the city (chapter 3), unintentionally, it has contributed considerably to groundwater recharge. Therefore it may be concluded that while the collected portion of domestic wastewater is reused for agricultural irrigation, on-site disposal of domestic sewage is a major source of groundwater recharge.

To evaluate these two reuse pathways, the respective wastewater quantities and qualities need to be quantified. Since the collected portion of wastewater is used for irrigation, it is also important to understand and examine the physical setting (extent, cropping pattern, farmers perceptions, etc.) of wastewater irrigation. The three components of quantity, quality and agricultural practices of wastewater irrigation are discussed separately below.

#### *7.7.2.1 Wastewater quantity*

In general, assessing wastewater quantities takes place through direct measurements of influent to treatment plants. In case of partial coverage of the sewerage system, knowledge of the water use can result in adequate estimation of wastewater quantities. In the Sana'a case, not only is the exact coverage of the sewerage system unknown (no regular measurements), the per capita water consumption is estimated to vary widely between those connected to the public water supply and those supplied by private vendors. Thus, using the available population and public water supply production data, generated wastewater quantities in the city can only be estimated (see chapter 3 for assumptions and calculations used to estimate wastewater disposal to on-site cesspits).

Historical data of wastewater flow of the sewerage system were collected from various reports. In order to obtain updated flow records, direct measurement of the water level above the inlet weir of the oxidation ponds were conducted in the period of 4-24 March 1997 by means of a 5 bar Tirtaharapan Absolute Pressure Logger.

In order to measure the total wastewater flow, arrangements were taken to block the bypass outlet so as to force the entire flow to pass the weir and thus enter the collection sump of the pond system. The logger was set to record water level readings with a frequency of once every five minutes. Even though the measurement period (20 days) does not indicate any seasonal variation, it took place during the spring, and hence could represent the annual average of the higher flows during the summer, as a result of the higher domestic water consumption, and lower flows during the winter.

#### *7.7.2.2 Wastewater quality*

Although, through on-site disposal, the infiltrating sewage undergoes quality improvements from its passage along the subsoil layers, quantification of these purifying processes and the assessment of the final quality are out of the scope of this research. Nevertheless, it has been concluded that groundwater contamination under the densely populated zones of the city is the result of extensive use of cesspits (chapter 3). In any case, it can be generalized that the quality of domestic raw sewage to cesspits is similar to that of the influent of the oxidation ponds.

Even though some quality data of the collected portion of wastewater were obtained from the various reports, in order to assess the suitability of wastewater for reuse, it was decided to collect updated samples along the effluent channel for biological and chemical analysis. Accordingly six sampling points were selected to include the inlet and effluent of the oxidation ponds in addition to four points along the effluent channel in the reuse region (Figure A2.1, Appendix 2). Based on their relevance to the different water quality guidelines (WHO, FAO and USEPA), different parameters were selected for analysis (Table 7.7).

Table 7.7 Selected parameters for wastewater analysis.

| **Parameter** | **Reference guidelines** | **Effects** |
|---|---|---|
| Total / Fecal coliform | WHO | Health |
| EC | FAO | Salinity |
| COD / BOD | - | Sewage strength |
| $Na^+$, B, $Cl^-$ | FAO | Toxicity |
| $Na^+$, $Ca^{++}$, $Mg^{++}$ | FAO | Soil permeability |
| N, pH, $HCO_3^-$, $CO_3^{--}$ | FAO | Irrigation |
| Heavy metals | FAO, USEPA | Toxicity |

During the period from May 1996 to March 1997, six sampling sessions were carried out for chemical analysis, and four sessions for microbiological (total and fecal coliform) tests. Due to logistical and time limitations, only grab samples were collected for analysis; nevertheless, the sampling sites were selected where the flow is nearly completely mixed (turbulence spot or free fall). Samples from all sites were collected at 6-7 a.m. and transferred to the laboratory for analysis. Composite samples could not be collected since the laboratory use was only limited to official working hours (8 a.m.–1:30 p.m.). All chemical analyses and the coliform tests were carried out in the NWSA central laboratory. All samples were filtered prior to analysis, where duplicate measurements were performed. All sampling and analytical tests were conducted according to standard procedures (APHA, 1980). Some chemical parameters were analyzed using the HACH DR/2000 direct reading spectrophotometer, others were analyzed using the titrimetric or the flame-photometric methods. The membrane filtration method was used for both total and fecal coliform tests. Technical difficulties prevented testing for helminths. Results of the analysis are shown in Figures 7.3-7.7 showing the standard error with a confidence interval of the mean at 95%. Variation in the results indicate a range in quality which include both the analytical error and the actual differences in wastewater quality, which may be caused by irregular disposal of septage from sucked cesspits into the sewerage system and the frequent fluctuations of the bypass flow at the ponds. Whereas results of the raw sewage and the pond effluent are the mean values of only one sampling site each, values of the reuse region are the mean values of 5 and 4 sampling sites for biological and chemical analysis, respectively. This was done to represent the entire reuse region.

#### *7.7.2.3 Agricultural practices and farmer behavior*

Since most of the ponds effluent is directly used for irrigation, knowledge of the cultivated area and cropping pattern in addition to farmers' experience, perception and acceptance of reuse is required. In order to reach a reliable estimate of the total wastewater irrigated area, a field visit was undertaken to determine the boundaries of the cultivated region along the effluent channel. These boundaries were then plotted on an official 1:50,000 map of the region, to approximately measure the total cultivated area. The cropping pattern of the area was determined through direct observations and interviews with various farmers during the different visits to the area.

During three field visits to the reuse region, discussions and informal interviews with approximately 20 farmers revealed their experience and perceptions towards wastewater irrigation. Although the interviews were not based on a written questionnaire, the obtained information can be considered reliable since the gained information could be cross-checked through observations and interviews with different farmers during other visits. Although the sample size of the interviewed farmers is quite small, the obtained information is of a general nature that is not expected to differ widely between farmers in the reuse region.

## 7.8 Results and discussions

### 7.8.1 Groundwater recharge

Eighty percent of the city's population discharges around 13-14 $Mm^3/a$ of raw sewage through cesspits to the underground (chapter 3). This discharge contaminates groundwater, thus leading to possible health problems from the drinking of contaminated groundwater and from direct exposure to overflowing wastewater. Although some quality improvement (reduction of suspended solids, BOD and pathogens) is expected from the passage of the leachate through the soil, and from dilution with groundwater, nitrogen and bacteriological contamination have been detected in local shallow as well as deep wells (chapter 3; Wiesenekker, 1996; Al-Hamdi, 1997). Nevertheless, given the water scarcity in the area, this discharge has augmented considerable amounts of recharge water, representing 25-60% of the total average annual natural recharge to the regional aquifer (Figure 7.8).

In the household, groundwater, whether supplied from the public distribution network or from private vendors, is mostly used for non-potable purposes; most drinking water is obtained from bottled mineral water or containers sold by private filtration stations. The continuous growth of the city and the financial constraints to expand the sewerage system, suggest that the use of cesspits will continue to expand for the foreseeable future. It is, therefore, advisable to re-evaluate, and regulate, direct discharge of untreated wastewater to the underground in order to maximize groundwater recharge, while at the same time minimizing the environmental and health effects. The mandatory use of on-site pretreatment units, such as septic tanks, prior to final discharge to cesspits could enhance the quality of percolating wastewater through a reduction in suspended solids and BOD, and may, on the long run, prove attractive for consumers (lower rate of clogging, and thus shallower cheaper pits).

### 7.8.2 Direct irrigation of fodder and cereal crops

#### *7.8.2.1 Physical setting*

As of 1997, a total of 14,200 $m^3$/d or 5.2 $Mm^3$/a of raw sewage is collected from the city to the oxidation ponds at Al-Rowdda, north of Sana'a. Part of the collected flow (in the order of 40–50%) is treated, while the rest is bypassed to the ponds' discharging point in a natural channel along the Bani-Howat wadi (Figure A2.1, Appendix 2). During the winter months (non-irrigation season) the water reaches a natural evaporation/infiltration pond near the village of Bait Daghaish, about 23 km north of Sana'a. During the summer, the flow is fully utilized by the farmers downstream of the ponds. Wastewater irrigation is mainly taking place within the wadi, along a strip of approximately 80 meters wide and 13 km long to form a total cultivated area of around 100 hectares. This area would not be used to cultivate fodder and cereal crops if wastewater were not available, but most probably it would be converted to dry farming (cereals) or groundwater irrigation (cash crops). The wastewater irrigation is dominated by the cultivation of sorghum, wheat, and alfalfa, in addition to some onions, tomatoes and potatoes. Naturally growing grass along the wadi is mainly used for animal grazing. Farmers obtain wastewater either from gravity diversion of the main channel to the fields or through pumping, using small portable suction pumps, from small retention ponds that facilitate pumping and at the same time yield stable settled sludge that is then used as fertilizer or soil conditioner. Basin irrigation is the main irrigation method; nevertheless, some furrow irrigation was observed. No protective measures[87] seem to have been taken by farmers during wastewater irrigation; nearly all agriculture related activities are performed manually in similar ways irrespective of the origin of the irrigation water.

With the exception of some vegetable plots, cash crops, particularly qat and grapes, are seldom irrigated with wastewater. Farmers consistently reported that wastewater irrigation of qat and grapes is mostly negative. Unspecified yield reductions and a change in the taste of grapes were suspected by farmers to be the result of wastewater irrigation. Among farmers, wastewater is thought to "burn" qat trees, as a change in leaf color and yield has been experienced in the past.

Direct wastewater disposal is prohibited by Law[88] unless the discharging agency (in this case, NWSA) takes the necessary measures to remedy all adverse effects

[87] Any measure to avoid contact with wastewater such as wearing boots or gloves.
[88] Environmental Protection Law (Law No. 26, 1995).

associated with such discharge. Article 30 of the Environmental Protection Law establishes regulatory standards for the collection, treatment, discharge and reuse of wastewater. These standards combine the three major guidelines of WHO, FAO, and USEPA for wastewater reuse (Ayers and Westcot, 1985; Mara and Cairncross, 1989; Pescod, 1992). The WHO guidelines were adapted to minimize the direct health hazards of wastewater irrigation. The FAO guidelines deal mainly with the chemical quality requirements for irrigation water and the USEPA recommendation limit trace elements in irrigation water.

#### *7.8.2.2 Wastewater quantities*

The total inflow to the ponds has been measured at several occasions. NWSA records indicate an average inflow of around 5,000 $m^3$/d in 1988. The continuous increase in house connections, both legal and illegal, has increased the flow considerably. With about 13,000 house connections, Suleiman (1990) estimated the influent to the ponds in 1990 to be 8,300 $m^3$/d, covering 17% of the city's population. Ludwig (1991) reported an estimate of 8,000 $m^3$/d of collected raw wastewater from Sana'a. In 1992, the total wastewater flow was measured to range between 8,000-8500 $m^3$/d (Veenstra et. al, 1995). The High Water Council (HWC, 1992) indicated that the total wastewater flow to the ponds was 8,000 $m^3$/d in 1990 and projected an increase to 13,200 and 26,400 $m^3$/d by the year 1995 and 2025, respectively. The total flow was measured for a period of 10 days in March 1994 to average 10,197 $m^3$/d, with an average daily peak factor of 1.86 (Howard Humphreys, 1995). Within the framework of this research, flow measurements show an average flow of around 14,200 $m^3$/d, with an average daily peak factor of around 1.5 (Figure 7.2).

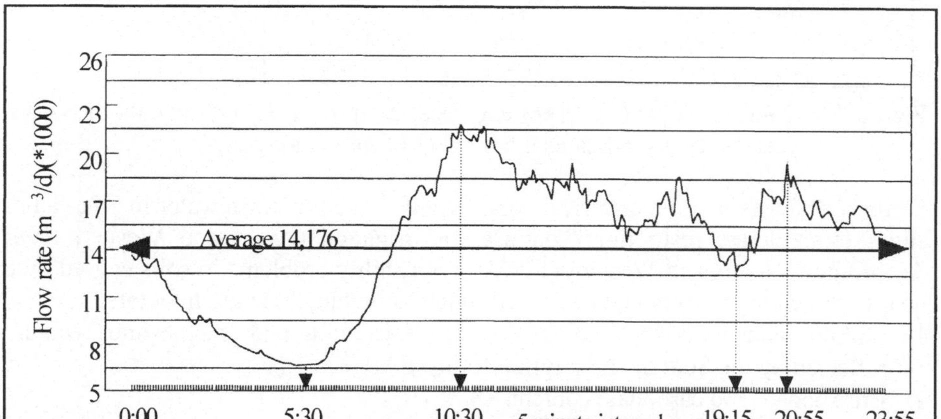


Figure 7.2 Average daily flow of wastewater to the oxidation ponds (4-24 March 1997).

#### *7.8.2.3 Wastewater quality*

Table A2.4 (Appendix 2) summarizes the results of historical routine analysis carried out by the NWSA laboratory. Table A2.5 (Appendix 2) shows the results obtained by Veenstra et. al. (1995). In the preparation for the design of the new activated sludge treatment plant, the strength of raw sewage was determined through sampling and analysis of the influent to the ponds (Table A2.6, Appendix 2) (Howard Humphreys, 1995). These results are reasonably consistent when compared with results obtained during this research. The high BOD (400-570 mg/l) and COD (750-1400 mg/l) values are indicative of strong wastewater that may be attributed to a low per capita water consumption. Within the framework of this research additional chemical and biological analyses of raw wastewater, treated wastewater (pond effluent) and channel effluent along the reuse areas were carried out. Figures 7.3-7.7 show the average quality of wastewater to be evaluated in accordance with the WHO, FAO and USEPA guidelines.

Figure 7.3 shows the removal ratio for total and fecal coliforms by the oxidation ponds to be 98.7% and 98.4%, respectively, whereas $10^3$-$10^5$ removal is required. When compared with the WHO guidelines, the remaining high fecal coliform counts in the effluent make it unfit for unrestricted irrigation. Moreover, the high ratio of bypassed raw sewage increases the coliform counts back to very high levels in the reuse areas. Given the high strength of wastewater and the inadequate treatment, it is suspected that wastewater in the reuse region does not meet the WHO helminth guidelines either.

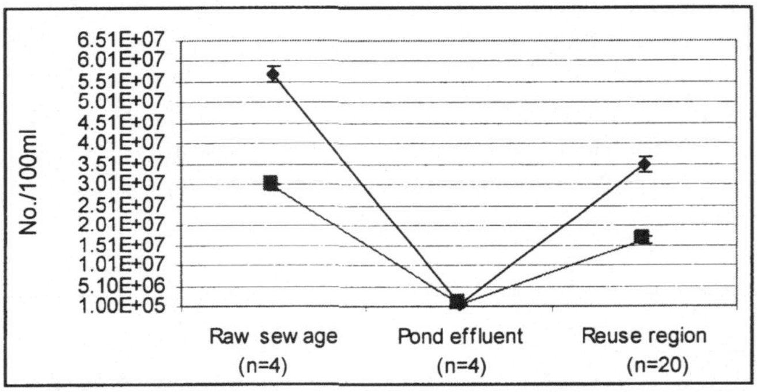


Figure 7.3 Total coliform (TC) (◆) and fecal coliform (FC) (■) counts of Sana'a's wastewater (n indicates the number of samples).

Figure 7.4 shows that the electrical conductivity[89] (EC) of wastewater in Sana'a has a slight to moderate restriction[90] on use for irrigation (Ayers and Westcot, 1985). According to Ayers and Westcot (1985), permeability problems in soils caused by the quality of irrigation water can be related to the following three main factors:

1. Sodium concentration relative to that of calcium and magnesium, which is reflected by the Sodium Adsorption Ratio (SAR)[91].
2. Bicarbonate and carbonate content.
3. The total salt concentration represented by the EC.

---

[89] EC of water reflects the amounts of dissolved salts, which if present in high amounts can accumulate in the crop root zone and reduce the water uptake ability of the plant to the extent that yield is affected (Ayer and Nestcot, 1985).

[90] A restriction on use indicates that there may be a limitation in choice of crop, or special management may be needed to maintain full production capability. A restriction on use does not indicate that the water is unsuitable for use (Ayer and Nestcot, 1985).

[91] $SAR = \frac{Na}{\sqrt{\frac{Ca + Mg}{2}}}; (concentrations\ in\ meq/l)$

The results suggests that no permeability problems are expected from the use of raw or treated wastewater from Sana'a for irrigation.

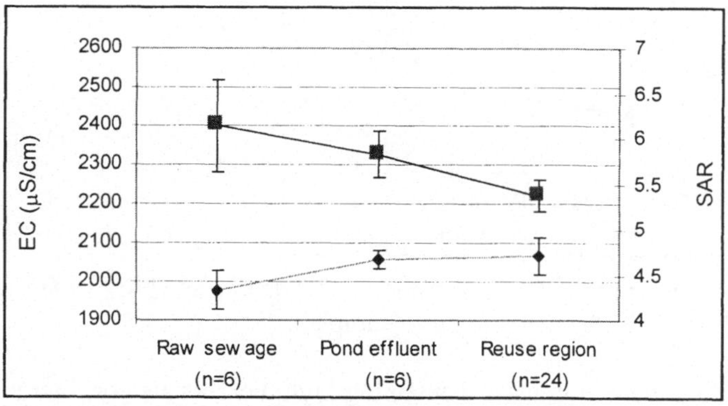


Figure 7.4 Salinity (EC) (■) and Sodium Adsorption Ratio (SAR) (◆) of wastewater in Sana'a (n indicates the number of samples).

Toxicity problems associated with irrigation water result usually in yield reduction and are associated with the presence of, notably boron, sodium and chloride (Ayers and Westcot, 1985). With a range of 0.9–1.5 mg B/l in the wastewater, the FAO guidelines recommend a slight to moderate restriction on the use for irrigation. Similarly, Figure 7.5 shows that sodium and chloride concentrations are also present at levels where slight to moderate restrictions are recommended.

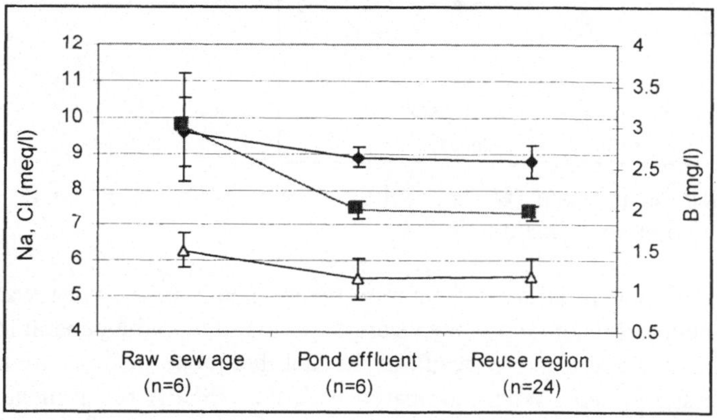


Figure 7.5 Toxic constituents (sodium (◆), chloride (■) and boron (▲)) of wastewater in Sana'a (n indicates the number of samples).

Although nitrogen application, either in the soluble form with irrigation water, or through fertilizer application, enhances crop productivity, high nitrogen application results in excessive vegetative growth, yield reduction, lodging and a delay in crop maturity (Ayers and Westcot, 1985). As can be seen from Figure 7.6, the wastewater from Sana'a contains small concentration of nitrate, but ammonia is present at levels that can cause severe problems especially to nitrogen sensitive crops.

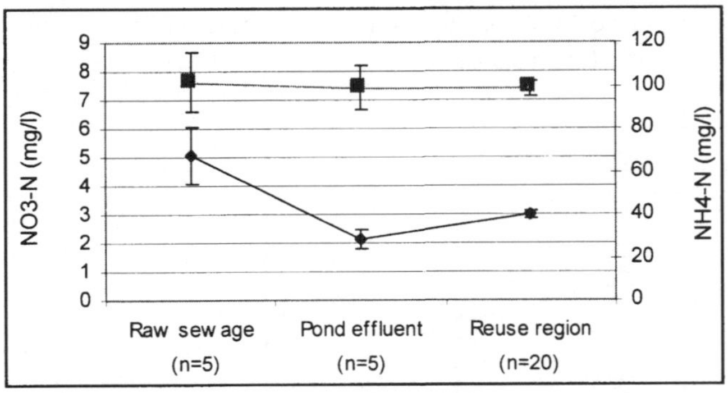


Figure 7.6 Nitrogen concentration in wastewater from Sana'a, as ($NO_3^-$) (◆) and ($NH_4^+$) (■) (n indicates the number of samples).

Using overhead sprinkling, irrigation water containing high bicarbonate concentration causes white deposits on leaves and fruits. Wastewater from Sana'a contains high bicarbonate content (>518.5 mg/l), which would create severe problems if sprinklers were used for irrigation (Figure 7.7).

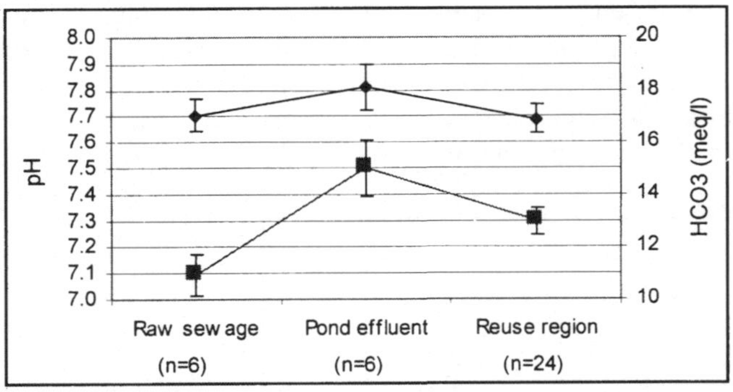


Figure 7.7 pH (◆) and bicarbonate (■) concentration of wastewater in Sana'a (n indicates the number of samples).

Table 7.8 shows average concentrations of common heavy metals in the wastewater from Sana'a. Low concentrations of heavy metals are expected, due to the general low industrial activity in the area and also as a result of the fact that most factories are not connected to the sewerage system. When compared with the USEPA recommended limits for irrigation water, both cadmium and chromium are present at levels that fall between the short and long term recommended limits. Copper, iron, lead and zinc, however, are present in concentrations below the recommended limits and thus do not pose any threat. Although heavy metals seem not to pose a major constraint to wastewater irrigation, a more comprehensive soil investigation in the area should indicate any limitations to the reported values.

Table 7.8 Average concentrations of heavy metals in wastewater from Sana'a in the reuse area.

| Element | Sana'a wastewater (mg/l) | USEPA recommended limits (mg/l) | |
|---|---|---|---|
| | | Long-term use* | Short-term use** |
| Cadmium (Cd) | 0.025 | 0.01 | 0.05 |
| Chromium (Cr) | 0.20 | 0.1 | 1.0 |
| Copper (Cu) | 0.13 | 0.2 | 5.0 |
| Iron (Fe) | 0.63 | 5.0 | 20.0 |
| Lead (Pd) | 0.25 | 5.0 | 10.0 |
| Zinc (Zn) | 0.188 | 2.0 | 10.0 |

* For water used continuously on all soils.
** For water used for a period of up to 20 years on fine textured natural or alkaline soils.

While Table A2.7 (Appendix 2) shows the results of the biological and chemical analysis conducted during this research, Table 7.9 summarizes the overall wastewater quality in Sana'a in relation to the WHO, FAO and USEPA guidelines.

Table 7.9 Summary of wastewater quality as compared to the WHO, FAO and USEPA guidelines.

| Parameter | Restriction degree on irrigation use: None | Slight to moderate | Severe | Wastewater quality in reuse region (Sana'a) | Remarks |
|---|---|---|---|---|---|
| **Salinity (FAO)** | | | | | |
| Ecw ds/m | <0.7 | 0.7-3.0 | >3.0 | 2.2 | Slight to moderate salinity. Can be used for salt tolerant crops |
| **Permeability (FAO)** | Ecw = | | | | |
| SAR =0-3 | >0.7 | 0.7-0.2 | <0.2 | SAR = 4.7 | No problem expected |
| =3-6 | >1.2 | 1.2-0.3 | <0.3 | | |
| =6-12 | >1.9 | 1.9-0.5 | <0.5 | | |
| =12-20 | >2.9 | 2.9-1.3 | <1.3 | | |
| =20-40 | >5.0 | 5.0-2.9 | <2.9 | | |
| **Toxicity (FAO)** | | | | | Moderate levels of toxicity. Expected to increase with continuous use. Tolerant crops should be selected. |
| Sodium (SAR) | <3 | 3-9 | >9 | 4.7 | |
| Chloride (meq/l) | <4 | 4-10 | >10 | 7.2 | |
| Boron (mg/l) | <0.7 | 0.7-3.0 | >3.0 | 1.2 | |
| **Miscellaneous effects (FAO)** | | | | | Severe problems are expected from high nitrogen. Removal is recommended or selection of tolerant crops. Sprinkler irrigation should be avoided to minimize effects of high bicarbonate. |
| Nitrogen ($NO_3$-N +$NH_4$-N) | <5 | 5-30 | >30 | 102 | |
| Bicarbonate (meq/l) | <1.5 | 1.5-8.5 | >8.5 | 13.0 | |
| pH | 6.5-8.4 | | | 7.7 | |
| **Health hazards (WHO)** | **Unrestricted irrigation** | **Restricted irrigation** | | | Unfit for unrestricted irrigation and high possibility not to be fit for restricted irrigation |
| Fecal coliform (No./100ml) | <1000 | - | | $18*10^6$ | |
| Intestinal nematode | ≤ 1/liter | ≤ 1/liter | | Not tested | |
| **Trace elements (USEPA)** | **Long-term use** | **Short-term use** | | | A broader investigation is needed for verification and sources identification especially for Cadmium and Chromium. |
| Cadmium (mg/l) | 0.01 | 0.05 | | 0.025 | |
| Chromium (mg/l) | 0.1 | 1.0 | | 0.2 | |
| Copper (mg/l) | 0.2 | 5.0 | | 0.13 | |
| Iron (mg/l) | 5.0 | 20.0 | | 0.63 | |
| Lead (mg/l) | 5.0 | 10.0 | | 0.25 | |
| Zinc (mg/l) | 2.0 | 10.0 | | 0.19 | |

### 7.8.3 Wastewater reuse in the context of water resources management

A tentative water balance of the Tawilah sandstone aquifer (Figure 7.8) puts the reuse potential in perspective by showing the relative quantities of available wastewater. Even though the maximal availability of wastewater represents only 8.5-9.5% of the total water use in the Sana'a basin area, it represents around 70% of the net domestic water consumption within the Sana'a city. Furthermore, the currently collected portion is only in the order of 15% of the total domestic water use, as the larger portion is discharged directly to the underground.

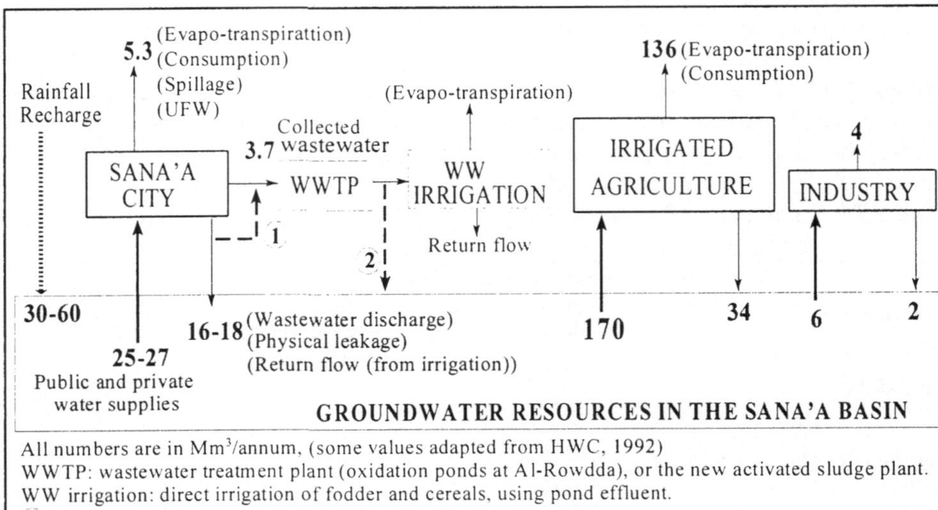


Figure 7.8 Tentative water balance for groundwater in the Sana'a basin (situation 1995).

Recently, funds have been allocated to construct a wastewater treatment plant to replace the existing over-loaded oxidation ponds. An extended aeration treatment plant has been designed with a hydraulic capacity of 50,000 $m^3$/d and an effluent of 30 mg BOD/l and 30 mg/l of suspended solids. The treatment plant is to consist of screens, detritors (grit removal), aeration, final sedimentation, chlorination and a polishing lagoon[92], in addition to sludge thickeners and drying beds. In response to the growing pressure from residents of areas in the city facing rapid clogging of cesspits, the government may decide to expand the sewerage system to cover those areas. After the recent coverage of Al-Hasabah quarter (1997), the remaining two parts, namely Nokom and Al-Lakamah represent around 10-20% of the city's population. Therefore, the sewerage system may expand in the near future to collect a total of 17,500-23,300 $m^3$/d, which would cover around 30-40% of the 1995 city's population, or around 18-25% of the projected population in 2010. Thus, it can be concluded that unless the sewerage system is further expanded to cover the entire city, the new treatment plant will probably operate below design capacity for a long time. The under-loaded conditions coupled with the 215,000 $m^3$-polishing lagoon suggest a

[92] Using the measured average flow of 14,200 $m^3$/d, this represents around 15 days of hydraulic retention time

better than designed effluent quality will be achieved. Given this relatively high effluent quality and the alarming extent of groundwater mining in the area, wastewater should be used to partially alleviate the pressure on the local groundwater resources.

## 7.9 Discussion of options for wastewater reuse in Sana'a

### 7.9.1 Identification of alternatives and criteria

Given the wastewater quantity and quality expected from the new treatment plant in Sana'a, six alternative reuse options can be identified for evaluation. These alternative *scenarios* are: (1) direct irrigation of fodder and cereal crops, (2) direct irrigation of existing groundwater irrigated cash crops fields, (3) groundwater recharge and recovery for irrigation of existing groundwater irrigated cash crop fields, (4) groundwater recharge and recovery for potable use, (5) municipal irrigation of public areas and green belts, and (6) non-potable domestic use. For simplicity these options will be referred to as scenarios (1) through (6) in the following sections. These alternatives are evaluated against five main *criteria* or *objectives*, i.e. (1) minimal health hazards, (2) minimal costs, (3) maximal public acceptance, (4) minimal environmental impact, and (5) maximal water resources conservation. These objectives are refered to as criteria (1) through (5). The identified scenarios are evaluated and accordingly scaled and valued against the five criteria factors (sections 9.2-9.6), and the outcome is summarized in Table 7.10. The valuation is based on a scale of 4 points, from –3 to 0, corresponding to increasing attainment of the objective.

### 7.9.2 Health hazards

The health effects associated with the presence of organic and inorganic chemicals in wastewater are excluded from this analysis due to the limited knowledge of their short and long-term health implications and the confined scope of this research. The analysis here is limited to the direct hazards from pathogenic organisms (helminths, bacteria and viruses). A scale of 0 to -3 is used, where 0 indicates no risk, while -1, -2 and -3 indicate low, moderate and high risk, respectively. For scenario 1, a low risk score -1 reflects the limited contact of wastewater to farmers and the disinfected nature of the effluent. Scenario 2, carries the additional potential risk to consumers, thus increasing the risk score to -2. For scenario 3, the associated health risk is almost nil due to the additional natural purification and dilution that takes place in the underground layers, which justifies a risk score of 0. If the recharged wastewater is to be recovered for potable use (scenario 4), then two factors influence the associated health risks, namely, the extra treatment prior to recharge, and the additional treatment prior to consumption. Nevertheless, given the origin of the recharged water, a low risk score of -1 is used in this analysis. For scenario 5, the main health risks will depend mainly on the reliability of the disinfiction process and the effectiveness of the management and regulation. Although it is quite difficult to assess the health risks associated with this alternative, especially given the inadequate local institutional environment, an overall low/moderate -1/-2 risk score is used. The health risk associated with scenario 6 depends on the extra treatment needed to upgrade the effluent to acceptable quality. The high contact level to which consumers are exposed, which includes washing, bathing, cooking, etc., justifies the use of a moderate risk score of -2.

### 7.9.3 Costs

Although for the purpose of comparison here the cost of sewage collection and treatment are excluded, the costs for the additional treatment, conveyance, injection, recovery and storage are to be included. A cost score of 0 to -3 is used to analyze each of the alternative options for additional costs, where 0 indicates no added costs, and -1, -2 and -3 reflect low, moderate and high additional costs, respectively. For scenario 1, no additional costs are considered since treated wastewater would flow through the wadi to be used by farmers along the natural effluent channel, and thus a score of 0 is used. For scenario 2, a cost score of -2 is used, which reflects the need for conveyance structures to transport treated wastewater to groundwater-irrigated cash crop areas, in addition to costs for inter-seasonal storage. For scenario 3, a cost score of -2 reflects the need for injection and recovery wells, in addition to conveyance structure to transport the recovered water to existing groundwater-irrigated cash crop areas. A maximum cost score of -3 is used for scenario 4, which reflects the need for extra treatment prior to recharge and delivery in addition to injection and recovery wells. In order to reuse the treated wastewater for scenario 5, a distribution system and storage facility are needed, reflected in a cost score of -2. For scenario 6, a dual distribution may be needed in addition to extra treatment and storage facilities, thus justifying the use of the maximum cost score of -3.

### 7.9.4 Public acceptance

In order to evaluate the reuse options with respect to public acceptance, the degree of bodily contact either to the farmer (occupational risk) or end-user is taken as a key variable, assuming that acceptance is inversely proportional to the perceived degree of bodily contact. A scale of 0 to -3 is used to quantify public acceptance, where a score of 0 indicates no contact and thus maximum acceptance, while -3 indicates maximum contact (drinking) and hence maximum rejection. Scenario 1 suggests a low level of bodily contact (limited to farmers) and since most cultivated crops are not intended for human consumption, an acceptance score of 0 is used. On the other hand, a score of -2 for scenario 2 is justified by the high likelihood that consumer preference would lower the value of effluent-irrigated cash crops, especially since freshwater-irrigated cash crops are widely available. Consumers may not object to cash crops irrigated with groundwater, originally recharged with treated effluent (scenario 3), and, thus, this alternative comes with a score of 0. On the other hand, even though extra treatment prior to both recharge and consumption may be necessary, public rejection is expected to be high if the recovered groundwater (scenario 4) is used for potable use. A maximum score of -3 is used in this case. Scenario 5 is associated with a limited degree of bodily contact, especially if precautions[93] are taken, thus reducing the score to a low -1. Higher degree of bodily contact associated with wastewater reuse for scenario 6 suggests the use of the higher score of -2.

### 7.9.5 Environmental impact

Even though a comprehensive assessment of the environmental impact of wastewater reuse is beyond the scope of this research, the direct effect of the reuse alternatives on the quality and integrity of groundwater is used as the main indicator of environmental impact. A score of 0 to -3 is used to evaluate the extent of groundwater

[93] Could include: awareness programs, warning signs, low pressure irrigation techniques, and nighttime irrigation scheduling.

contamination; 0 indicates no damage and -3 reflects maximum contamination and irreversible deterioration. For the reuse alternatives that involve irrigation of fodder, cereals, cash crops, or municipal parks (scenarios 1, 2, 5), a score of -1 is used to reflect the effect of return flows[94] on the quality of the receiving groundwater. For scenario 3, a maximum score of -3 reflects the expected local long-term change in the quality of groundwater. The extra wastewater treatment prior to the recharge and recovery for potable use (scenario 4) justifies the low risk score of -1. Since no direct discharge is associated with scenario 6, a score of 0 is appropriate.

### 7.9.6 Conservation of water resources

Reuse of wastewater has an additional, intrinsic economic and financial value by delaying or lowering the need for capital investment for the development of more expensive water resources. Even though the concept of water conservation is usually included within the broader issue of environmental impact, discussed in the previous section, within the context of this discussion, environmental impact was confined to quantitative impact of wastewater reuse, whereas water conservation is discussed from a qualitative viewpoint. Given the seriousness of water shortage in Sana'a, the reuse of treated wastewater should contribute to a delay in the need for expensive water supply alternatives like water import and sea water desalination. To evaluate the different reuse alternatives, a score of 0 to -3 is used. A conservation score of 0 indicates a maximum contribution towards water conservation and commensurate reduction in groundwater abstraction, and -3 indicates no contribution. Although the use of treated wastewater for scenario 1 is not expected to reduce groundwater abstractions, a small percentage of return flow may recharge the underlying aquifers, thus justifying a low conservation score of -2. The same score is used for scenario 5 and is justified by the small amounts of return flow and the present absence of groundwater irrigation of municipal parks. A maximum contribution towards water resources conservation, 0, is expected from scenarios 2, 3, 4 and 6.

### 7.9.7 Scenario evaluation and selection

During the previous analysis, cross-valuation between those variables was not taken into account. In other words, the conservation score of -3, which indicates a lack of contribution towards water resources conservation, is not necessarily equal in weight to the -3 factor for public acceptance. Even though it is beyond the scope of this work, partial cross-valuation of the five variables is attempted below.

The first four criteria are directly related to the quality of the wastewater, while only the last criterion is related to its quantity. Because of the overriding importance of the water scarcity, this criterion should be attributed more weight by multiplying all conservation factors by a common correction factor. This correction factor can be approximated through the ratio between the cost to develop a sustainable new water resource, which in Sana'a's case is import of desalinized seawater, and the cost to upgrade wastewater to high quality through advanced treatment. It is reported that desalination and delivery of desalinized seawater to Sana'a from the coastal area would cost around 7.5 US$/m$^3$, while secondary plus advanced wastewater treatment[95] costs are in the order of 2-2.5 US$/m$^3$ (Asano, 1985; Kruseman, 1997). The ratio of the two costs, 3, is used to correct the factor values (Table 7.10).

---

[94] The portion of irrigation water that infiltrates and reaches groundwater.

[95] Including reverse osmosis.

Table 7.10 Multi-criteria analysis of wastewater reuse scenarios in Sana'a.

| Criterion / Scenario | (1) | (2) | (3) | (4)[a] | (5)[b] | Arithmetic sum[b] |
|---|---|---|---|---|---|---|
| (1) | -1 | 0 | 0 | -1 | -2 (-6) | -4 (-8) |
| (2)[c] | -2 | -2 | -2 | -1 | 0 (0) | -7 (-7) |
| (3)[c] | 0 | -2 | 0 | -3 | 0 (0) | -5 (-5) |
| (4) | -1 | -3 | -3 | -1 | 0 (0) | -8 (-8) |
| (5) | -1/-2 | -2 | -1 | -1 | -2 (-6) | -7/-8 (-11/-12) |
| (6) | -2 | -3 | -2 | 0 | 0 (0) | -7 (-7) |
| | 0 = No risk<br>-3= Max. risk | 0=No extra cost<br>-3=Max. cost | 0=Max. acceptance<br>-3=Min. acceptance | 0=No threat<br>-3=Max. threat | 0=Max. contribution<br>-3=No contribution | |

a Reflects mainly groundwater contamination.
b Values between brackets have been multiplied by a correction factor of 3.
c Using wastewater to replace groundwater.

Table 7.10 identifies two distinctive categories of feasible reuse alternatives: those with a summation of less than –5, and those with a summation of greater than or equal to -5. The two scenarios 1 and 3 have the highest overall score and are the most appealing. On the other hand, scenarios 5 and 4 have the lowest potential. Even though wastewater reuse for fodder and cereal crops appeared earlier to be an obvious option, it shifts to the less-appealing category after corrected for water scarcity. Although this alternative brings an extra income for some local farmers, it does not contribute towards the conservation of water resources. It is clear from Table 7.10 that the only truly attractive alternative for wastewater reuse in Sana'a is groundwater recharge and recovery for irrigation of existing groundwater-irrigated cash crop fields.

### 7.9.8 Sensitivity analysis

There are three main variables involved in the previous analysis, namely the type of reuse scenario, the assessment criteria, and the water scarcity correction factor. Though more alternatives can be identified for reuse such as for industrial or recreational uses, the selected options are believed to be of higher priority and sufficiently representative. Similarly, more assessment criteria could be added to specify in more detail the concepts of, e.g., health and environmental impacts, but it is believed that such more detailed analysis would be of a benefit in a separate more comprehensive study. The same argument can also be used to justify the method used to weigh the different scenarios.

A closer look at the valuation of the different scenarios reveals that the criteria 3.4 and 5, (public acceptance, environmental impact and contribution to the conservation of water resources) were reasonably accurately valued according to the coupled conditions of bodily contact for criterion 3, the quantity of wastewater to percolate and recharge groundwater for criterion 4, and the quantity of groundwater to be freed from agriculture for criterion 5. Nevertheless, no clear conditions were used to justify the valuation of Criteria 1 and 2 (health hazards and costs). This is mainly due to the lack of available reliable data on the conditions that can be used for such valuation e.g., outbreaks of water born and water related diseases and their associated causes and the energy requirements for the different additional treatment units associated with each scenario. Therefore, it is believed that a higher potential for error exists in the valuation of the health factors and to a larger extent the cost factors. To reduce the uncertainty associated with the valuation of these two criteria factors, the following two checks are performed:

1. Assume that an overestimation of the cost factor for a particular scenario has occurred while underestimation of the same factor has taken place for the remaining scenarios. To correct for this, a new valuation of the scenarios is repeated six times, where for each round the cost factor for a particular scenario is decreased by one point and increased by one point for the remaining scenarios (Table 7.11).

Table 7.11 Outcome of valuation under overestimation of cost factor by one point for one scenario and underestimation of cost factors by one point for each of the remaining scenarios.

| **Corrected scenario** | **Algebraic sum*** | **Scenario with lowest algebraic sum** |
|---|---|---|
| (1) | -4 (-8) | Sc. 3, with a sum of –6. |
| (2) | -6 (-6) | Sc. 2, 3 with a sum of -6 |
| (3) | -4 (-4) | Sc. 3, with a sum of -4 |
| (4) | -7 (-7) | Sc. 3, with a sum of -6 |
| (5) | -6/-7 (-10/-11) | Sc. 3, with a sum of –6 |
| (6) | -6 (-6) | Sc. 3, 6 with a sum of -6 |

* Sum of overall factors (value between parentheses are corrected sum for water scarcity)

Table 7.11 shows scenario 3 to the most attractive reuse option. When scenarios 2, or 6 are revalued, their overall score match the original score of scenario 3. In practical terms this means that, if indeed an overestimation of the cost factor by one point for scenarios 2, or 6, while an underestimation of the cost factor by one point for scenario 3 has occurred during the original valuation, the scenario in consideration (2, or 6) becomes an equally viable option similar to that of scenario 3. Moreover, scenario 1 is still an unfeasible option and only becomes attractive when the water scarcity factor drops to below 1.5.

2. Assume that an overestimation of both the health and cost factors for a particular scenario have occurred, while these factors for the remaining scenarios stay at the original valuing shown in Table 7.10. Under this assumption, revaluation of the different scenarios was repeated six times, where for every time the health and cost factors for a particular scenario is decreased by one point each, with no change done to the remaining scenarios (Table 7.12).

   Again, Table 7.12 shows scenario 3 to be the most attractive option. When the health and cost factors of scenarios 2, or 6 are overestimated by one point each under the original valuation procedures, the overall score of the scenario under consideration (2, or 6) would match the original overall score of scenario 3, and thus becomes an equally attractive option. Scenario 1 on the other hand becomes an attractive alternative only when the water scarcity factor drops to below 2.0.

Table 7.12 Outcome of valuation under overestimation of the health and cost factors by one point each for one scenario and underestimation of the health and cost factors by one point each for each of the remaining scenarios.

| **Corrected scenario** | **Algebraic sum*** | **Scenario with lowest algebraic sum** |
|---|---|---|
| (1) | -3 (-7) | Sc. 3, with a sum of –5. |
| (2) | -5 (-5) | Sc. 2, 3 with a sum of –5 |
| (3) | -4 (-4) | Sc. 3, with a sum of -4 |
| (4) | -6 (-6) | Sc. 3, with a sum of –5 |
| (5) | -5/-6 (-9/-10) | Sc. 3, with a sum of –5 |
| (6) | -6 (-6) | Sc. 3, 6 with a sum of -5 |

* Sum of overall factors (value between parentheses are corrected sum for water scarcity)

It can therefore be summarized that under the presented assumptions[96] that underlie the previous analysis, the reuse of treated wastewater in Sana'a for groundwater recharge and recovery to replace groundwater irrigation of cash crops (scenario 3) is the most suitable alternative. As can be seen from the sensitivity analysis, under special conditions, reuse of treated wastewater for direct irrigation of cash crops as to replace groundwater (scenario 2), or for non-potable domestic uses (scenario 6) are shown to be viable options as well. In any case, the current practice of wastewater reuse for irrigation of fodder and cereal crops are shown to be unfeasible unless the water scarcity factor (the ratio of the costs to develop a reliable potable water source to the cost to upgrade wastewater to potable quality) drops to below 1.5. Even with the shortcomings of the valuation procedure, it can be argued that the analysis provides a reliable assessment of wastewater reuse in the context of integrated water resources management in the Sana'a region.

## 7.10 Conclusions and recommendations

Due to a low per capita water consumption in the Sana'a city, the wastewater production quantity is small (7-9%) when compared to over all water demand in the basin area.

As of 1997, the collected portion of wastewater totals around 5.2 $Mm^3/a$, which represent about 20-25% of the total wastewater production within Sana'a. This portion is currently being utilized by farmers, mostly to irrigate fodder and cereal crops along the effluent channel.

At present, 75-80% of the total raw sewage from the city is discharged to the underground via cesspits, thus artificially recharging the underground aquifers, but also causing unchecked local groundwater contamination. This contamination has lead a large portion of the population to rely on private unregulated small filtration stations for drinking water.

Even though direct irrigation of fodder and cereal crops is considered financially beneficial to a small number of farmers, it has not reduced any pressure on the local groundwater resources.

---

[96] Particularly those used to determine the water scarcity correction factor.

At present, the quality of the wastewater makes it unsuitable for agricultural reuse, which is mainly attributed to the unsatisfactory treatment in the overloaded oxidation ponds and high percentage of bypassed flow.

With the new treatment plant (under construction), there will be greater capacity to withstand future expansion of the sewerage system and produce adequate quality effluent for further reuse.

Given the local circumstances in the Sana'a basin, evaluation of six reuse alternatives shows that groundwater recharge and recovery for cash crops irrigation is likely the most appropriate alternative.

### 7.11 References

Abuzahrah B. (1995) Impact of groundwater pollution on water supply in Gaza Strip. Ms.C. thesis no. EE-177, IHE, Delft.

Al-A'ama M. and Nakhla G. (1995) Wastewater reuse in Jubail, Saudi Arabia. *Wat. Sci. Tech.* **29**(6), 1579-1584.

Alaerts G. J. (1996) Generalized analytic framework for institutional studies on water, IHE paper, Delft.

Alaerts G. Mahbubar Rahman M. and Kelderman P. (1996) Performance analysis of a full-scale duckweed-covered sewage lagoon. *Water Research.* **30**(4), 843-852.

Alaerts G., Blair T. and Hartvelt F. (1991) A strategy for water sector capacity building. IHE report series 24, Delft.

Al-Hamdi M. (1997) Case study XIII Sana'a, Yemen. In: Helmer R. and Hespanhol I. (eds.). *Water pollution control*, E&FN Spon, London.

Ali I. (1987) Wastewater criteria for irrigation in arid regions. *Journal of irrigation and Drainage engineering.* **113**(2), 173-183.

Arar A. (1991) Wastewater reuse for irrigation in the Near East region. *Wat. Sci. Tech.* **23**, 2127-2134.

Asano T. (ed.) (1985) *Artificial recharge of Groundwater.* Butterworth, Boston.

Asano T., Maeda M. and Takaki, M. (1996) Wastewater reclamation and reuse in Japan: Overview and implementation examples. *Wat. Sci. Tech.* **34**(11), 219-226.

Ayers R. and Westcot D. (1985) *Water quality for agriculture.* Irrigation and Drainage Paper 29. FAO, Rome, Italy.

Banks P. (1991) Wastewater reuse case studies in the Middle East, *Wat. Sci. Tech.* **23**, 2141-2148.

Bartone C. and Arlosoroff S. (1987) Irrigation reuse of pond effluents in developing countries. *Wat. Sci. Tech.* **19**(12), 289-297.

Bartone, C. (1994), Urban sanitation, sewerage and wastewater management: Responding to growing household and community demand. ESD, the World Bank, Washington D.C.

Bontoux J. and Coutois G. (1996) Wastewater reuse for irrigation in France. *Wat. Sci. Tech.* **33**(10-11), 45-49.

Bouwer H. (1992) Agricultural and municipal use of wastewater. *Wat. Sci. Tech.* **26**(26), 1583-1591.

Bruvold W. (1988) Public opinion on water reuse options. *Journal WPCF.* **60**(1), 45-49.

Farooq S. and Ansari Z. (1983) Wastewater reuse in Muslim countries: An Islamic perspective. *Environmental Management.* **7**(2), 119-123.

Haarhoff J. and Merwe B. (1996) Twenty five years of wastewater reclamation in Windhoek, Namibia. *Wat. Sci. Tech.* **33**(10-11), 25-35.

Howard Humphreys (1995) Design Report and review of Impacts. National Water and Sanitation Authority, Sana'a, Yemen.

HWC (1992) Water Resources management Options in Sana'a Basin. Final Report Vol. 9, Technical Secretariat of the High Water Council, Sana'a, Yemen.

Kruseman G. (1997) SAWAS Exclusive Summary. SAWAS-3, National Water and Sanitation Authority, Sana'a, Yemen.

Ludwig H. (1990). *Report on appropriate sewage treatment for Sana'a*, National Water and Sanitation Authurity, Sana'a, Yemen.

Maeda M., Nakada K., Kawamoto K. and Ikeda M. (1996) Area-wide use of reclaimed water in Tokyo, Japan. Wat. Sci. Tech. **33**(10-11), 51-57.

Mara D.and Cairncross S. (1989) Guidelines for the safe use of wastewater and excreta in agriculture and aquaculture. WHO, Geneva.

Nair S. and Selvam P. (1996) India: Madras – Renovated sewage for agriculture, industry and groundwater recharge. Presented in the conference on Recycling waste for agriculture: The rural-urban connection. The World Bank, Washington D.C.

Pescod, M. (1992) *Wastewater treatment and use in agriculture*. Irrigation and Drainage Paper 47. FAO, Rome, Italy.

Postel S. (1992) *The Last Oasis-Facing Water Scarcity*. Earthscan Publications Ltd.

Shelef G. (1991) Wastewater reclamation and water resources management. *Wat. Sci. Tech.* **24**(9), 251-265.

Shelef G. and Azov Y. (1996) The coming era of intensive wastewater reuse in the Mediterranean region. *Wat. Sci. Tech.* **33**(10-11), 115-125.

Shuval H. (1991) The development of health guidelines for wastewater reclamation. *Wat. Sci. Tech.* **24**(7), 149-155.

Shuval H., Adin A., Fattal B., Rawitz E. and Yekutiel P. (1986) Wastewater Irrigation in Developing Countries: Health Effects and Technical Solutions. World Bank Technical Paper No. 51. The World Bank, Washington D.C.

Skillicorn P., Spira W. and Journey W. (1993) Duckweed aquaculture: A new aquatic farming system for developing countries. The World Bank, Washington D.C.

*Standard Methods for the examination of water and wastewater* (1980). $15^{th}$ edn, American public health association, New York, USA.

Suleiman M. (1990) Wastewater, treatment and re-use in the Yemen Republic. UNDP/DTCD project YE/88/001, Technical Secretariat of the High Water Council, Sana'a, Yemen.

UN (1985) *Wastewater reuse and its application in Western Asia*. Economic and Social Commission for Western Asia, E/ESCWA/NR/84/2/Rev.1.

UN (1997) Critical Trends: Global Changes and Sustainable Development. United Nations Publications, New York.

USEPA (1992) Manual guidelines for water reuse. U.S. Environmental Protection Agency, Washington D.C.

Veenstra S., Al-Nozaily F. and Alaerts G. (1993) Purple non-sulfur bacteria and their influence on waste stabilization pond performance in the Yemen Republic. *Wat. Sci. Tech.* **31**(12). 141-149.

Walsh M. (1996) Salalah residents to benefit from treated wastewater. *World Water and Environmental Engineering*. March issue.

Wiesenekker H. (1996) Microbiological Groundwater Contamination of the Sana'a City Area. Technical note No. 23, SAWAS-3, National Water and Sanitation Authority, Sana'a, Yemen.

Wilkinson J. (1978) Islamic water law with special reference to oasis settlements. *Journal of Arid Environments*. **1**, 87-96.

# Chapter 8

# Conclusions and Recommendations

## 8.1 Overexploitation and consequences

As a result of the exclusive reliance on groundwater use for municipal and agricultural uses, the main challenge to the Sana'a basin is the on-going groundwater mining (2-6 m/a drop in water levels). The current total groundwater abstraction, of which irrigation accounts for around 83%, is in the order of 6 times that of the natural recharge. The causes for the increase in groundwater use are correlated with the high growth rate of the urban population (annually 8-10% since the 1970s) and the sustained high profits generated by the cultivation of cash crops. Water availability and the profitability of cash crops have changed the cropping pattern in the Sana'a basin, converting large areas under traditional rain harvesting land into qat cultivation, and uncultivated land and land under cereals into grape production. Farmers in the basin area behave rationally, but are strongly risk-avert in crop choice, agronomy and irrigation habits. Risk-avert, here, means that farmers' decisions are led by short-term perspectives and maximization of net income at the expense of longer-term sustainability of water resources. The high net return from qat and grape cultivation (8,000-13,000 US$/ha for at and 3,000-7,000 US$/ha for grapes) is believed to be a strong incentive for farmers to reduce the risk of possible lower yields through application of access irrigation water (1.3-1.6 m/a for at and 0.6-1.3 m/a for grapes) resulting always in over-irrigation of qat[97].

In order to control and better manage groundwater abstractions, an institutional change is needed. Such change can come about through external intervention in the form of, for example direct state control, or internal intervention, through the establishment of self-organized institutions to deal with the problem. In both cases, the issue of water ownership and water use rights is critical. Water use rights should be clearly defined, which requires the establishment and enforcement of both boundary and authority rules. Boundary rules, through certain criteria, clarify and identify access to groundwater, thus limiting the number of pumpers. In practical terms, among other consequences, this would mean a stop in further drilling. The establishment of authority rules will reinforce the boundary rules through determining the criteria upon which water rights are distributed. Example criteria of boundary rules might be land ownership, well shareholding, endogenous status of the pumper, etc.. Criteria of authority rules might be historical abstraction patterns, size of irrigated land, cropping pattern, or a combination of more than one criterion.

Failure to initiate an institutional change (external or internal), means that the current trend in groundwater use will continue at an increasing rate, leading to near total

[97] Over-irrigation of qat is a common practice including areas with acute water shortage through the purchases of supplemented irrigation water.

depletion of the aquifers. The adoption of this no-intervention policy would result in the following consequences:

- Gradual breakdown of the economic base in most of the rural area of the basin, as sub-basins on the plain will run dry one after the other, leading to a relatively high local unemployment.
- A partial conversion to dry farming practices that would induce a shift in land use and cropping pattern, returning to pre-1970 practices, partially mitigating the expected increase in unemployment.
- A continuous price rise and possible decline of the coverage of the public water supply in the urban area. The resulting water shortages would pressure the government to implement costly water supply development projects, thus exerting unwelcome financial pressure on the government as well as on the city dwellers. Higher financial burden on the city dwellers resulting from an increase of the share of water and sanitation services in a typical household budget from 1-7% for those connected to the NWSA supply and 3-13% for those relying on private tanker supply, to around 6-17% for imported water from Surdud or Marib, and to 18-49% for imported desalinized sea water[98].
- Greater dependence on private domestic supplies leading to higher rate of abstractions from private wells in the city, thus inducing higher decline in water levels, and/or faster recycling of wastewater leachate from under the city.
- Higher domestic water prices would force the poor to rely on comparatively unsafe "contaminated" water for drinking leading to higher risk of outbreaks of water borne and water related diseases.

## 8.2 Water use rights and local institutions

Although officially all natural resources are considered state property, the absence of a detailed legal framework, the lack of institutional setup to administer and manage water resources and the weak capacity of the government to devise and enforce appropriate water related policies, have allowed the local users to overexploit many of the fossil aquifers. In practice, groundwater rights in the Sana'a basin seem to be (1) connected to the ownership of the overlaying land, and (2) governed by the financial capacity of the landowner, often in cooperation with neighboring farmers, to access the aquifer. This has resulted in a rapid expansion in drilling within the basin area from a few wells in the early 1970s to more than 5,000 boreholes by the mid 1990s, mostly for irrigation. To overcome the problem of land fragmentation and the increasing drilling costs, well sharing provided a pragmatic solution to access a reliable irrigation water source with the shareholders contributing to the capital and recurrent costs proportional to their shares. Well sharing provides a solid framework for equity and economic efficiency through a reduction of the financial burden on the farmers, and higher utilization rate of the abstraction equipment. With partial backing of the religious *shari'ah* principles, the current "informal" water right system seems to be widely adopted and unchallenged by the government. Nevertheless, although the

[98] The variation in the share of water and sanitation in the household budget varies according to the type of the service and the income level. Although the NWSA supply is subsidized, to calculate the share of water and sanitation for the proposed supply projects, no subsidy for the water supply was assumed.

priority[99] in water use within the *shari'ah* principles is satisfied on the small scale of the vicinity of the well, it is unwittingly violated at the larger, basin-wide scale.

Although the local social tribal structure may be considered an obstacle that hinders government control over groundwater, especially when groundwater is the critical input for the farmers' livelihood, the tribal structure should be considered to play an effective role in a joint management concept of groundwater in the basin area. The results of the study display the tribe as a coherent socio-political institution with consistent potential capacity for water allocation and conflict resolution, and identify Sheiks (both at the village and tribe levels) as influential figures in collective decision-making and conflict resolution.

## 8.3 Groundwater use, economic efficiency, equity and sustainability

It has been argued that in order to achieve economic efficiency in the use of water resources, water should be viewed as an economic good (the Dublin Declaration, 1992). In other words, water is to be used in the highest value use, which in many cases suggests reallocation of water from agriculture to municipal and industrial uses. In order for such reallocation to take place, administrative regulation and/or competitive water markets are to be established as to facilitate the transfer of tradable water rights to higher value uses. Tradability of water rights suggests at least partial separation of water rights from the land ownership. Failure to achieve such reallocation means economic inefficiency and a high opportunity cost for the "wasted" water, particularly in water scarce regions. In order for this concept to be reflected in the Sana'a situation, the uses of groundwater in the industrial, municipal, and agricultural sectors need to be evaluated and compared. Due to the insignificant water demand of the industrial sector relative to that of the other two sectors, the value of groundwater use in industry is not further discussed. Within the Sana'a city, groundwater is used by both NWSA and private vendors to satisfy the domestic water need of the urban population, which perceives the water service as unreliable both in quantitative and qualitative terms. Because of the high subsidies on the NWSA supply, tariffs are so low that they do not even recover the operation and maintenance costs, and in real terms have been lowering since the early 1990s, therefore, the NWSA supply is not included in the comparison. Given the small but high-value quantities of needed drinking water, purchases from filtration stations would probably continue even after reallocation of groundwater from agriculture to domestic use justifying their exclusion from the comparison. This means that the value of water in the private non-potable domestic use is to be compared to that of groundwater irrigation. Within irrigated agriculture, groundwater is mainly used for the cultivation of cash crops, particularly qat, grapes and vegetables. Since qat and grapes are the largest groundwater using crops in the basin (around 65%), the value of water use for these two crops will be used to represent groundwater irrigation. The value of water use will be reflected from the net financial return obtained from its use.

Chapters 2 and 6 show that while water use for private domestic supply provides a net return of 15-40 and 60-94 YR/$m^3$ for piped and tanker supplies, respectively, the net return from water use in qat and grape cultivation is 70-110 and 40-95 YR/$m^3$, respectively. Financial return from groundwater use in qat and grape cultivation is

[99] In the Shari'ah principles, priority in water use is given first to human drinking followed by animal drinking and irrigation.

therefore higher than that of the domestic supply despite the relatively high tariff of the private domestic tanker supply (1.2-1.5 US$/m$^3$). In other words, it may be concluded that groundwater use in the Sana'a basin is economically efficient, despite the relatively undefined water rights and the absence of a competitive market. Factors that have helped achieve such economic efficiency in groundwater use are:

- The high demand for qat and grapes in the city, which is partially due to the rapid expansion and growing wealth of the city during the last three decades.
- The relatively easy access to groundwater, with a reasonably "unchallenged" water right system (attachment to land and governed by financial capacity).
- The reliability of groundwater as an irrigation water source and the financial capacity of farmers to overcome the decline in water levels.
- The limitation to viable cropping patterns, where it has become financially unfeasible to cultivate alternative crops, e.g., cereals.

Given the growing water shortage in the area, suppose that the value of water use in the municipal sector increases and surpasses that of qat cultivation, thus justifying a shift in water use to urban domestic supply, the large difference between the water quantities used in each sector raises an important discrepancy in the opportunity cost. While the opportunity cost of the quantity needed for the domestic sector may be high, having satisfied that need by reallocation from irrigated agriculture the opportunity cost of the remaining quantity in the irrigation sector drops to almost zero, hence justifying its use for lower return uses. This is particularly true since no boundary and authority rules exist that limit groundwater withdrawals to levels that would prevent such discrepancy in opportunity costs.

Thus, it may be concluded that the outcome of the current situation in the Sana'a basin is similar to that of a competitive market with tradable water rights, resulting in reasonable economic efficiency, productivity and, given the relatively easy access to groundwater through well sharing, an acceptable level of equity. Nevertheless, due to the absence of boundary and authority rules to limit the withdrawable quantity, in Sana'a, such gains are at the expense of resource sustainability. In other words, assuming infinite storage of the aquifer, or high level of recharge that exceeds abstractions, the existing situation in the Sana'a basin, with regard to water use and assuming qat production supports the economy, would be socially attractive and economically close to its optimum.

With the high vulnerability of groundwater in the Sana'a basin, which threatens the welfare of the urban and to a larger extent the rural population, groundwater use in the basin should not be valued on the basis of the existing uses, but rather on the scarcity value of the resource. In other words, groundwater in sana'a should be valued on the basis of the replacement, or marginal costs. Recognizing that water import projects would be too costly to justify their use for agriculture, the current no-intervention policy can be viewed as a subsidy to the rural population that the government and/or the urban population would have to pay for in the future through water import for domestic supply.

## 8.4 Incentive systems of groundwater use in Sana'a

With reference to the analytical framework of this research illustrated in Figure 1.3 (Chapter 1), the incentive system for groundwater use, both in irrigation and urban domestic use is accordingly expressed in Figure 8.1. As for groundwater use in irrigation, it is clear that the financial incentive, resulting from the relatively high profits of qat and grapes cultivation has probably been the main driving force behind the observed expansion in irrigated agriculture over the past three decades. A second indirect, but equally important incentive is the current ambiguous legal status of water ownership, displaying opposing views between the government (constitution) and the *shari'ah* principles. The current situation of uncontrolled groundwater mining is further accelerated by the low capacity of the government to monitor or effectively enforce the permit system[100] introduced in 1973. It seems that some farmers are aware of the deplorable nature of the supplying aquifer(s) and in effect are taking advantage of the otherwise lost opportunity. It is also believed that loosing the current groundwater resource would mean a lose of irrigated agriculture since there aren't adequate alternatives that can satisfy the current level of agricultural water demand not to mention their prohibitive costs.

With qat and grapes dominating the cropping pattern, it is predictable to observe a strategy on water use dominated by wasteful practices. Cash crops and qat in particular is often over-irrigated to compensate for labor and/or reduce the risk of inadequate yield. With no real strong effort to challenge the farmers' use of groundwater, from a sustainability stand point, farmers are abusing the resource for private gains thus creating a typical common-pool-resource dilemma. With large individual gains, the large size of the basin and the complexity of the groundwater system, the lack of reliable information, the heterogeneity of the pumpers and the general mistrust between, on one hand the tribes and the government and on the other hand among the tribes of the basin, it is expected that the user (farmers) would be unwilling to resolve the on-going competition over groundwater.

The official concern over groundwater mining in the Sana'a basin has been mainly linked to the urban domestic water supply especially given the prohibitive costs of the alternative water supply schemes. With high growth of the urban population since the 1970s, the coverage of private water supplies has been increasing to compensate for the inability of the public supply to maintain or expand its coverage, which is the result of low investment and inadequate cost recovery. For the poor, this translates into higher water bill and a possible shift to questionable quality water for drinking.

[100] The permit system is viewed as the only attempt by the government to challenge the on-going informal water use right system.

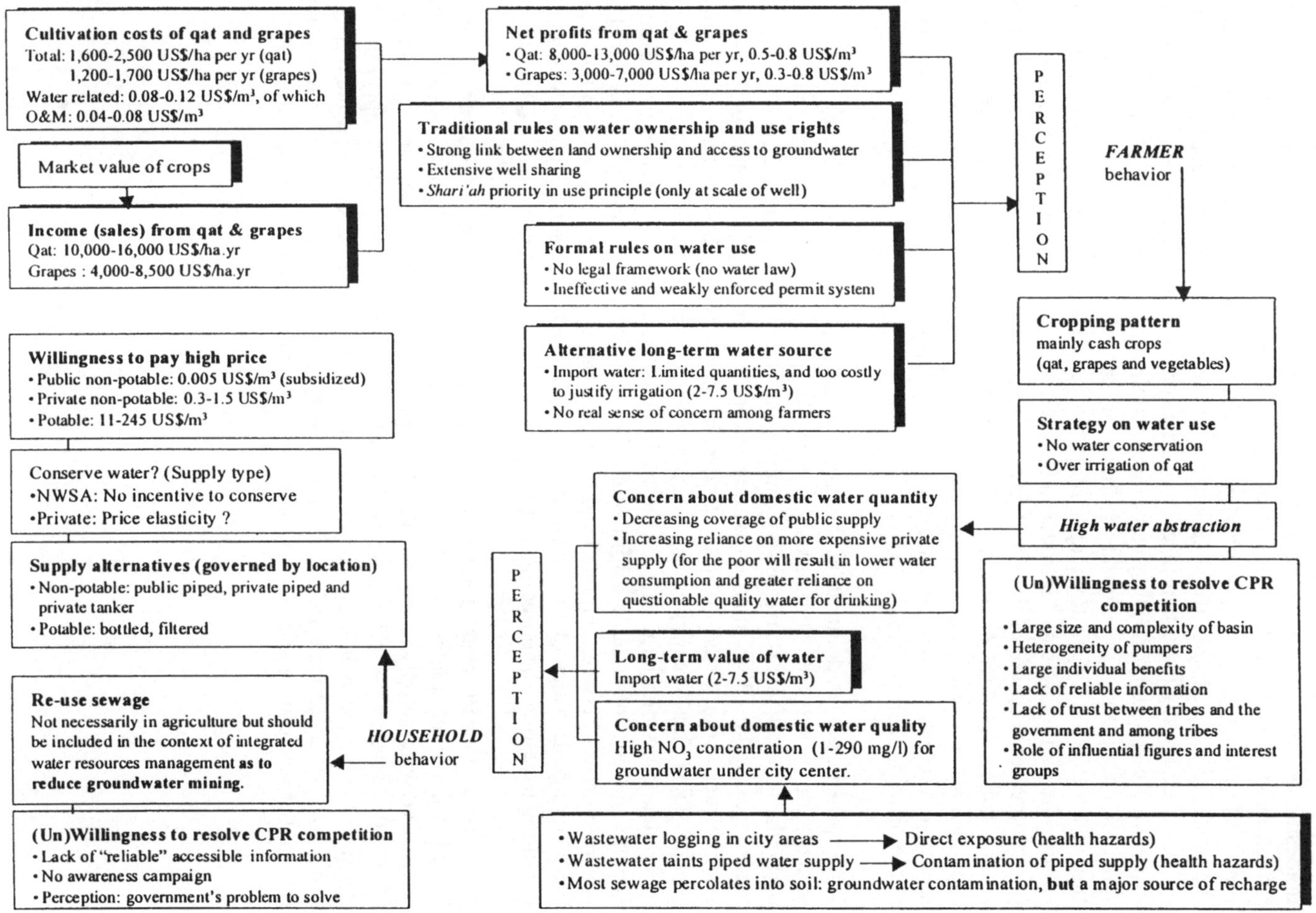


Figure 8.1 The incentive systems for groundwater use in irrigated agriculture and urban domestic supply (shaded: major incentives).

Concern over groundwater quality is directly linked to the extensive on-site disposal of domestic sewage to the underground via cesspits. Given the limited availability of reliable data on the complex hydrogeological setting of the basin, it is difficult to identify with certainty the degree and extent of the current contamination. Nevertheless, it is clear that many private deep boreholes, used for domestic supply, have been identified to contain elevated concentrations of nitrate and high levels of coliforms. In addition to groundwater contamination, on-site disposal of wastewater in areas with low infiltration capacity of the soil creates two additional health concerns from, (1) exposure to overflowing wastewater, and (2) the possible contamination of the public piped water supply[101]. Despite the deterioration in the quality of groundwater, which has lead to the popularity and expansion of small private filtration stations for drinking purposes, on-site disposal of wastewater has been a major source of groundwater recharge thus sustaining water levels under the city.

The type and level of water supply and consequently the price and share of water in the household budget in Sana'a seems to be governed by location of the household. With large price variation between the different supply alternatives, poor households relying on private tanker supply for non-potable water and filtered drinking water from supermarkets are the most effected. Therefore, it can be concluded that the decision to conserve water is tied to the type of supply, where the subsidized low tariff of the NWSA supply is viewed to be a deterrent for water conservation. Although it is suspected that the higher price of the private supplies is an incentive for wiser and thus lower water consumption, more research into the price elasticity of domestic water supply under water stress situations is needed.

Given the current acute water shortage and the expected growing scarcity in the "very" near future, domestic wastewater seems to be a reliable and growing water source that should not be wasted in low value uses, e.g., irrigation of fodder and cereal crops. More attention should be given to the management of treated wastewater so as to alleviate some pressure off, and, thus, replace groundwater for higher value uses, e.g., groundwater recharge, domestic non-potable uses, or irrigation of cash crops. Finally, although the city's dwellers seem to pay much attention to the day-to-day water problem on its "micro" scale (e.g., where to obtain the next tanker load) they seem not to be concerned with the overall gloomy "macro" scale of the problem. It is believed that the confusing and unreliable information and the lack of awareness reflect a perception that the water problem, at least on its large-scale, is the government's problem to solve.

## 8.5 Issues for improved groundwater management

It is unlikely that farmers would initiate institutional change due to their lucrative benefits that accrue from the current situation. It is important, therefore, that the government is involved to initiate the steps to reach better joint management of the available groundwater. The government should play a role different from that of an "exclusive" owner or controller. Nevertheless, the government's involvement should be strong enough to devise, supported by the participation of the local population, boundary rules to identify clear criteria for access to the resource, and most critically, for drilling. The government should also assist the local communities to devise their

[101] This takes place under the conditions of wastewater ponding and large pressure fluctuations (intermittent supply).

authority (distributional) rules to reach sustainable utilization of the aquifers, or more realistically, to achieve controlled groundwater mining. The objective of such rules is to arrest further expansion in groundwater use and to try to gradually reverse the current process of over-exploitation. The potential likelihood for success of this scenario increases if the basin area is subdivided into smaller, more socially (tribal) and politically coherent regions (not necessarily on hydrogeological basis) and if those groups can be motivated to initiate the establishment of water institutions through various incentives e.g., better extension services and subsidized water saving technology (such as drip irrigation). Such schemes should first be experimented in pilot regions, where groundwater is experiencing very rapid decline in water level (pronounced and visible externalities) and where the users form a homogenous group, so as to create a success experience for other regions to follow. The government should not rely on the existing official local administrative setup to deal with the water problem, but rather deal directly (for example through the National Water Resources Authority (NWRA)) with the local farmers themselves and try to utilize the capabilities of influential figures within the local communities. It is quite critical that such influential persons (mostly the local Sheiks) understand the extent of the problem and realize the collective benefits for themselves, the local farmers and their future generation. The government should also be involved in a basin-wide major communication and awareness campaign to convey more accurate information regarding the deteriorating state of groundwater and the expected consequences under the different scenarios.

Having devised the boundary and authority rules and thus having defined water use rights, negotiation between the city and the local rural communities can start to transfer such rights from irrigation to urban domestic use. With clear and defined water rights, such negotiations should not be based on land acquisition, but can be based on the transfer of the rights to groundwater while farmers retain the land and convert to alternative irrigation or farming practices. Compensation from such transfers can be used to develop the local community, to develop surface water resources to compensate for the reallocated groundwater, to invest in water-saving irrigation techniques, or it can be simply divided between the individuals proportional to their transferred rights.

Given the large subsidies associated with groundwater use, it has been suggested that the adoption of an appropriate pricing policy to increase the input costs of groundwater pumping would result in a substantial water savings, particularly in the irrigation sector. Diesel is the main energy source for groundwater pumping in the Sana'a basin with an average consumption of about 0.5 liter per $m^3$ of pumped groundwater. With heavy groundwater use, the subsidy on diesel exerts a substantial financial burden on the government. The low cost-profit ratio in qat and grape cultivation (15-25% for qat and 25-45% for grapes) undermines any effort to influence groundwater use through an increase in the input costs. The 67% rise of diesel price at the end of 1997 resulted in only a slight increase of the fuel cost component in the water-related costs, namely from 28% in 1997 to 35% in 1998. Increasing the input cost by doubling the diesel price would increase the cost-profit ratio of qat by only 5%. Assuming constant profit levels and adopting the unrealistic scenario of doubling the price of fuel, maintenance, and capital components, would increase the cost-profit ratio of qat by around 20%. This strongly suggests that measures to increase the input costs would most likely not lead to lower water application on qat, but would probably lead to a further shift in cropping pattern increasing the land under qat.

Although initiation of institutional change is a difficult task, the government should consider dealing with water scarcity from an integrated perspective. In areas where there is a potential for surface water development, i.e. rainwater harvesting, wastewater availability, availability of spring flow, etc., an integrated approach becomes feasible to influence groundwater use. The positive experience with the construction of the Shahik Dam in the Khowlan district, which influenced groundwater availability by reversing the decline in groundwater levels, is an example where the governmental technical expertise and intervention are useful. The entire basin area should be surveyed to identify possible appropriate recharge sites for similar projects. Community participation, and support, during all steps of these projects become essential to ensure a high level of institutional sustainability.

With the expected improved effluent quality from the new wastewater treatment plant, wastewater should be utilized to contribute to better water resources management. In a water scarce region like the Sana'a basin, the use of good quality effluent should not be wasted in a low value use like fodder and cereal cultivation. The government should engage in awareness programs in these areas so that farmers understand the value of treated wastewater as a reliable long-term water source. Incentives should also be created to encourage farmers to use wastewater in return for a reduction in groundwater abstractions. Such incentives may take the form of better extension services to deal with the agronomic problems associated with wastewater irrigation and the establishment of senior rights over wastewater to those willing to reduce groundwater pumping, and use wastewater for high value uses. Although such restrictions on wastewater use may sound unequitable, it is a well-targeted and effective method to reduce some groundwater withdrawals. Such process is expected to be more fruitful with further expansion in the sewerage system. Another and arguably more effective alternative is the use of wastewater effluent for groundwater recharge and recovery for irrigation of cash crops. This alternative has several advantages in that the aquifer serves as a storage facility during the non-irrigation season, breaks the psychological barrier to use wastewater, provides further water quality enhancement, and may eliminate or reduce the need for a costly distribution system.

Although the total wastewater production in the city of Sana'a is rather low when compared to the overall groundwater abstraction in the Sana'a basin (in the order of 7-9%), unlike groundwater, wastewater is a water source of growing volume with long-term reliability. Of the total wastewater production in the city, the sewerage system currently covers only 20-25% of the urban population. This collected portion is partially treated and informally reused for irrigation, while the remaining part of the city's domestic sewage is discharged to the underground via on-site cesspits. Although the quality of the collected wastewater is currently unsuitable for irrigation, it is illegally used for direct irrigation of fodder and cereal crops along the discharge wadi. Nevertheless, substantial quality improvements are expected when the new treatment plant enters operation in mid-2000. Extensive use of cesspits has lead to a substantial recharge of the aquifers, but has also resulted in the contamination of groundwater under the densely populated zones of the city. Given the high reliance of over half of the urban population on private wells within the city for their water needs, the contamination threat has lead many people to rely on private unmonitored filtration station for drinking water. The main quality concern of water from these filtration stations is related to the cleansing procedure of the reusable containers.

Assuming that the government deals with the threat of groundwater contamination by expanding the sewerage system to cover the entire city, the following consequences are expected:

- Higher rate of decline of groundwater levels under the city.
- A decrease in the coverage of the private domestic supplies as some private wells dry out.
- An increase in the price of private supplies so as to cover the cost of pumping the additional drawdown.
- With lower coverage of private supply, the higher water shortage will exert an additional pressure on the government to expedite the implementation of costly water import projects.

Meanwhile, if the collected sewage is treated and reused for fodder and cereal cultivation, it can be argued that wastewater reuse has shifted from the higher value use of non-potable domestic supply (after percolation through the subsoil layers and dilution with groundwater) to a lower use. Therefore, unless the collected wastewater is utilized for a high value use, e.g., cultivation of cash crop, it is "wasted" and an economic inefficiency occurs. Nevertheless, city zones experiencing wastewater ponding from clogged cesspits should be sewered by priority to minimize the health hazards from direct exposure to the overflowing wastewater, as well as to reduce the likelihood of water contamination in the NWSA distribution network that results from intermittent supply (pressure fluctuations) in those areas.

The complex geological setting of the Sana'a Basin will cause groundwater not to get exhausted in all sub-regions of the basin simultaneously. Rather, some sub-regions will dry up earlier than others. One after the other, farmers and other pumpers will relinquish water abstraction, which will gradually reduce the pressure on the resource. Conversion to rain-fed farming is expected, but many farmers would still be forced to cease agriculture altogether and seek other employment opportunities (notably, migration to Sana'a). There is insufficient hydrogeological and other data to predict how the gradual disconnection of irrigated agriculture will take place and to what extent it will achieve the reduction necessary to attain longer-term sustainable groundwater abstraction. Nevertheless, the outcome of the situation will certainly increase the social frictions, both in the rural and in the urban setting. In order for the government to minimize risks (such as migration, poverty, unemployment, etc.), to pre-empt subsequent shifts of other types of unsustainable groundwater mining in the basin, it is still necessary for the government to improve and adopt effective regulation of groundwater abstraction.

Given this complexity, this study has analyzed groundwater mining from different angles emphasizing the role of the incentive systems for both the urban domestic and the agricultural water use. However, to arrive at comprehensive strategies for sustainable water management in the basin it would be appropriate to address issues that are directly or indirectly related to (ground)water use, but were not explored, such as:

1. Voluntary population resettlement, which may be achieved through economic incentives, i.e. by shifting the main economic activities from water-short regions to areas where water resources are not so vulnerable. An obvious example is the relocation of the capital.
2. Introduction of appropriate sanitation technologies that better suit the aridity of the country. In this regard the feasibility of non-water borne sanitation in Yemen should be explored.
3. Lifting the import restrictions on qat. Qat cultivation accounts for the larger part of groundwater abstractions in the Sana'a basin (and probably also in other groundwater aquifers in the nation). Just like with many other agricultural products, imported substitutes could turn out cheaper. Hence, a social-political-economic analysis of such a policy should be conducted, as it may prove effective in reducing the pressure on the groundwater resources.

# Appendix 1.
## Questionnaire used during the field survey

Well Number:

Farmer's name:

Share in the supplying well:

Total area of farm:

Q1- Date of drilling? ___________.

Q2- Total depth? ______meter.

Q3- depth to water level after drilling?

Q4- Depth to water level now?

Q5- What was your source of money for drilling the well?
a( ) Private savings.
b( ) Borrowed from friend.
c( ) Loan from agriculture credit bank, with interest rate of___%.
d( ) Acquired free through a governmental project.
e( ) Other (specify)________________________________.

Q6- What was the purpose of drilling the well?
a( ) Irrigation.
b( ) Household use.
c( ) Water selling.
d( ) Other (specify) ____________________.
e( ) Combination of the above, (specify)__________.

Q7- Who is responsible for operating the well (incl. generator, pump, etc)?
a( ) Owner.
b( ) Co-owner with largest share.
c( ) Co-owner.
d( ) Hired person with a monthly salary of __________YR.
e( ) Other (specify)____________________.

Q8-What type of energy source for ground water abstraction?
a( )Electricity
b( ) Diesel generator
If diesel, capacity of the original installed generator?

Q9- Have you changed the generator since drilling? ( )Yes, ( )No
If Yes fill the following table:

| No. | date | Reason for change | Type of new generator | cost |
|---|---|---|---|---|
| | | | | |
| | | | | |

Q10- Have you changed the pump since drilling?( )Yes, ( )No
If Yes, fill the following table

| No. | date | Reason for change | Type of new pump | cost |
|---|---|---|---|---|
| | | | | |
| | | | | |

Q11- What is the yield of the well now? (to be measured In the field) ( liter/ sec)

Q12- Have you deepened the well since drilling? ( )Yes, ( )No
If (Yes) fill the following table

| Date | extra depth | Reason for deepening | cost |
|---|---|---|---|
| | | | |
| | | | |
| | | | |

Q13- If levels continue to decline will you deepen the well further?( )Yes, ( )No.

Q14- Do you sell water from your well on a regular basis?( )Yes, ( )No
If Yes fill the following table

| Buyer | Conveyance methodr | Tariff structure | Tariff (unit price) |
|---|---|---|---|
| | | | |
| | | | |

Q15- What crops do you cultivate in your farm (including rented)? (Area in Libnah=44m$^2$)
Grapes ( *area* ), Qat ( *area* ), Other ( *crop and area* )

| Crop | Area (Libnah) | In case rented annual rent | Crop type | | Reason for crop selection |
|---|---|---|---|---|---|
| | | | past | present | |
| | | | | | |
| | | | | | |
| | | | | | |

Q16-

| crop | # of irrigations last year | Average application time (plot area/hours) | Conveyance method | Irrigation method |
|---|---|---|---|---|
| | | | | |
| | | | | |
| | | | | |

Q17-

| Fertilizer use (Last year) | | | pesticide use (last year) | | | Hired labor (last year) | |
|---|---|---|---|---|---|---|---|
| Type | Amount | Cost | Type | Amount | cost | Number/days | cost |
| | | | | | | | |
| | | | | | | | |
| | | | | | | | |

Q18-

| Crop | Machinery use (last year) | | Taxes (last year) | | Sales | |
|---|---|---|---|---|---|---|
| | Type | cost | Amount | Harvest (amount) | Type | price |
| | | | | | | |
| | | | | | | |
| | | | | | | |

Q19- Normally speaking, what time of the year do you pump groundwater for irrigation?
Irrigation season (____________), Non-irrigation season (____________).

| Irrigation season | | Non-irrigation season | |
|---|---|---|---|
| Pumping period (hours/day) | Pumping period (days/month) | Pumping period (hours/day) | Pumping period (days/month) |
| | | | |

Q20- How much diesel did you buy for irrigation last year? Liter /barrel

| Diesel consumption ($l_{diesel}$/no. Of operation hours) | Diesel price |
|---|---|
| | |

Q21- How much lubricating oil did you buy for irrigation last year? Liter
Cost per liter? ( )

Q22- How much did you spend for maintenance last year?

| Date | Type of maintenance | cost |
|---|---|---|
| | | |
| | | |
| | | |

Q23- In case the power supply is electricity what was the electricity bill last year?

Q24- In general what do you think is your major expenses?
a( ) Diesel.
b( ) Oil.
c( ) maintenance.
d( ) Labor.
e( ) Fertilizer.
f( ) pesticides.
g( ) Other (specify)___________.

Q25- What can you say about your profits from farming for the last five years?
a( ) Have been increasing for the last five years.
b( ) Have remained the same for the last five years.
c( ) have continuously declined for the last five years.
d( ) Other (specify)____________________________.

Q26- Sometimes diesel shortages happen, How did you deal with diesel shortages?
a( ) I always store enough diesel during shortages so no effects on irrigation practices.
c( ) I applied less water to save diesel, nevertheless yield was not affected.
d( ) I applied less water to save diesel, but yield was affected.
e( ) Stopped irrigation.
f( ) Other (specify)____________________________.
If answer is (c or d) ask the following :
- How much water would you say was saved?
a( ) < 10%
b( ) 10 - 30%
c( ) 30 - 50%
d( ) > 50%
e( ) Other (specify)_________.
- After diesel became available, what did you do?
a( ) Changed back to original irrigation practice.
b( ) Continued the water saving measure.
c( ) Other (specify)______________________.

Q27- Do you think farmers will reduce their irrigation water use if diesel becomes more expensive?
( )Yes, ( )No.

Q28- How will your irrigation water use change if diesel becomes more expensive?
a( ) No change as long as farming is profitable.
b( ) No change, but will raise the price of crops.
c( ) change irrigation methods to pump less water and thus save diesel.
d( ) Other (specify)______________________________________.
e( ) Don't know what I will do.

Q29- As you know groundwater level in the area is declining, what would be your reaction if the government took certain actions to limit water pumping for irrigation?

| Governmental action | Farmer reaction |
|---|---|
| Charge for water | |
| Measure and limit the flow from well | |
| Raise the price of diesel | |

Q30- As you may know groundwater levels in the area have been declining since the 1980's, Who do you think is responsible for this decline?
a( ) NWSA.
b( ) Neighboring well owners.
c( ) Low rainfall.
d( ) Other (specify)______________________.

Q31- When you drilled the well, did you think that groundwater could be depleted?( )Yes, ( )No.

Q32- Do you now think that groundwater could be depleted?( )Yes, ( )No.

Q33- Has there been any land/water disputes (between you and your neighbor farmers) in the district?
( )Yes, ( )No.
If yes fill the following table :

| No. | Date | Type of dispute | Dispute description | Resolved | Who resolved | Period it took | How resolved |
|---|---|---|---|---|---|---|---|
| | | | | | | | |
| | | | | | | | |
| | | | | | | | |

Q34- who decides on the type of crops to be planted?
a( ) Head of household.
b( ) Son.
c( ) Sheik
d( ) Agricultural organization (specify)________________.
e( ) Other (specify)______________________.

Q35- who takes care of farm finances (sell crops, buy fertilizer, settle accounts etc.)?
a( ) Head of household.
b( ) Son.
c( ) other (specify)______________________.

Q36- who takes decisions regarding finances (e.g., investment in irrigation technology)?
a( ) Head of household.
b( ) Son.
c( ) other (specify)______________________.

Q37- Who do you refer to for religious matters?
a( ) Mosque Imam.
b( ) Shiek.
c( ) Other (specify)______________________.

Q38- What can you say about his knowledge of farming practices?
a( ) High, since he is a farmer.
b( ) High, but he is not a farmer.
c( ) Normal, similar to that of a non-farmer.
d( ) Low.

Q39- Do you think he could resolve conflicts related to water use?( )Yes, ( )No.

Q40- What can you say about the knowledge of the member of the parliament representing this district of farming practices?
a( ) High, since he is a farmer.
b( ) High, but he is not a farmer.
c( ) Normal, similar to that of a non-farmer.
d( ) Low.

Q41- Do you think he could resolve conflicts related to land or water?( )Yes, ( )No.

Q42- in general who is responsible to resolve water related conflects?
(1)______(2)______________________________
(3)______(4)______________________________

Q43- Who do you think owns groundwater?
a( ) Owner of over laying land (with restrictions)
b( ) Owner of over laying land (no restrictions)
c( ) The government
d( ) NWSA.
e( ) Only district residence.
f( ) The total community (incl. Sana'a city)
g( ) Other (specify)__________________________.

Q44- Do you think that you are careful (conservative) towards water use?( )Yes, ( )No.
If Yes How?

Q45- Do you think your neighboring well owners are careful towards water use?( )Yes, ( )No.
If Yes How?

Q46- Do you know of any regulations, restrictions or laws on groundwater pumping?
a( ) No.
b( ) 1973 protection law.
c( ) Other (specify)____________________.

Q47- Does one need a permit to drill a well?( )Yes, ( )No.

Q48- Do some people drill without a permit?( )Yes, ( )No.

Q49- In your opinion, what is the purpose of the permit?
A( ) A permit to drill only, and has nothing to do with the quantity to be abstracted
B( ) A permit to drill and also control the quantity to be abstracted
C( ) An income source to NWSA
C( ) Other (specify)

Q50- Did you get a permit to drill the well?( )Yes, ( )No
If answer is (yes) ask the following:
-How difficult was it to get the permit?
a( ) Not difficult if you know the right people.
b( ) Not difficult, but a lot of paper work.

Q51- Has any official checked your well in the last five years?( )Yes, ( )No
If answer is (yes) fill the following table:

| Official | Checked what | How often | Honest |
|---|---|---|---|
| | | | |
| | | | |
| | | | |

Q52-Is there a local non-governmental agricultural or otherwise organization? ( )Yes, ( )No
If Yes fill in the following table:

| | |
|---|---|
| Name of organization | |
| Head of organization | |
| Benefits to farmers | |
| Influence on crop selection | |
| Influence on water use | |
| Membership fee | |
| Role of farmer | |

***Wastewater reuse:***

Q1- Have some people from the district used or are using effluent from the ponds for irrigation?
( )Yes, ( )No.

Q2- Is it necessary to obtain a permit to use wastewater for irrigation?( )Yes, ( )No.
If answer is (yes) ask the following:
-What agency is responsible for issuing the permit?
a( ) NWSA.
b( ) Ministry of health.
c( ) Ministry of housing.
d( ) Ministry of agriculture.
e( ) Other (specify)__________________.

Q3- In your opinion, What are the dangers that are associated with wastewater irrigation?
a( ) Disease transmission to farmers through direct contact.
b( ) Disease transmission to consumers through crop consumption.
c( ) (a + b)
d( ) Other (specify)________________________________.
e( ) Don't know.

Q4- Have you ever used or are using wastewater for irrigation?( )Yes, ( )No.
If answer is (no) ask the following:
-If wastewater becomes accessible to you will you use it?( )Yes, ( )No.

-If answer to (Q4) is (yes) ask the following:
-Are you still doing so? ( )Yes, ( )No.
-When did you start and for how long did you irrigate with wastewater?_______.
-How did/do you get wastewater?
a( ) Pump it direct from the ponds to the field.
b( ) Pump it from the effluent channel to the field.
c( ) Pump it from ponds to tanker.
d( ) Pump it from effluent channel to tanker.
e( ) Other (specify)____________________________.
-How did you use wastewater for irrigation?
a( ) Irrigated only with wastewater.
b( ) Diluted wastewater with groundwater with ratio of (___________).
-What crops did/do irrigate with wastewater?
a( ) qat.
b( ) Grapes.
c( ) Vegetables.
d( ) Grains (specify)__________.
d( ) a combination (_______).
-What was/is the effects on crops and harvest?
a( ) Positive (higher yield), So far no negative effects.
b( ) Positive in the beginning but negative later.
c( ) Negative (lower yield) from the start.
d( ) No effects.

-What was/is the effects on the soil?
a( ) Positive (more productive), So far no negative effects.
b( ) Positive in the beginning but negative later (specify)_________.
c( ) Negative from the start (specify)____________________ .
d( ) No effects.
-Was/is it cheaper to irrigate with wastewater than groundwater or buying water?
a( ) No change in cost between the two.
b( ) 3/4 the cost of groundwater irrigation.
c( ) between 1/2 - 3/4 of the cost of groundwater irrigation.
d( ) Between 1/4 - 1/2 of the cost of groundwater irrigation.
e( ) < 1/4 of the cost of groundwater irrigation.
-What precautions did/do you take during irrigation with wastewater?
a( ) None.
b( ) Avoid contact as much as possible, but no physical precautions.
c( ) Long/special boots.
d( ) Gloves.
e( ) combination (specify) (_______)
f( ) Other (specify)________________________.

Q5- If, due to continuous water level decline, your well goes dry or groundwater becomes very expensive to buy, will you use wastewater for irrigation? ( )Yes, ( )No.
If answer is (no) ask the following:
-What will you do?
a( ) Change to dry farming.
b( ) Stop agriculture.
c( ) Other (specify)___________________________________.
d( ) Don't know.
If answer to (Q5) is (yes) ask the following question:
-Will you be willing to pay for wastewater?( )Yes, ( )No.
If answer is (yes) ask the following:
-How much will you be willing to pay?
a( ) Less than present groundwater extraction costs/price.
b( ) The same as the present groundwater extraction costs/price.
c( ) More than the present groundwater extraction costs/price as long as farming is profitable.

Q6-In your opinion, from the religious point of view, Is it acceptable to use wastewater for irrigation?
( )Yes, ( )No.

Q7- In your opinion, Is the effluent from the existing ponds treated well enough to be used for irrigation?( )Yes, ( )No.
If answer is (no) ask the following:
-As you may know a new (more advanced) treatment plant will be constructed soon, when this treatment plant is ready and if the effluent is declared safe to be used for irrigation, do you think you will use it for irrigation? ( )Yes, ( )No.

Q8- Would you feel more confident to use wastewater if the government demonstrated the short/long term effects of wastewater irrigation on state owned pilot farms? ( ) Yes, ( )No.

Q9- Should the government send people to farmers to explain the dangers of wastewater irrigation?
( )Yes, ( )No.

Q10- Do you think the government should pay farmers to use wastewater for irrigation?( )Yes, ( )No.
If answer is (yes) ask the following:
-Why?
1)__________________________________________.
2)__________________________________________.
3)__________________________________________.

Q11- If you decide to use wastewater for irrigation, How will your present water use be affected?

a( ) What ever wastewater I use, I will reduce the same amount from groundwater.

b( ) I will not reduce the amount of groundwater, and wastewater will be used to expand the cultivated area.

c( ) Other (specify)________________________________________.

Q12- In your opinion, using wastewater for irrigation without groundwater dilution, will:

a( ) Increase salinity of the soil.

b( ) be toxic to the soil.

c( ) Have no affects on the soil.

d( ) Other effect (specify)_____________________________________.

e( ) Don't know.

Q13- What consequences do you predict by using wastewater to irrigate qat, grapes and vegetables?

| Crop | Consequences | | | | | |
|---|---|---|---|---|---|---|
| | Product price | Growth | taste | yield | Disease transmission | Other |
| Qat | | | | | | |
| Grapes | | | | | | |
| Vegetables | | | | | | |

# Appendix 2.
Tables and Figures related to wastewater quality and reuse in Sana'a

Figure A2.1 Wastewater reuse areas in Sana'a.

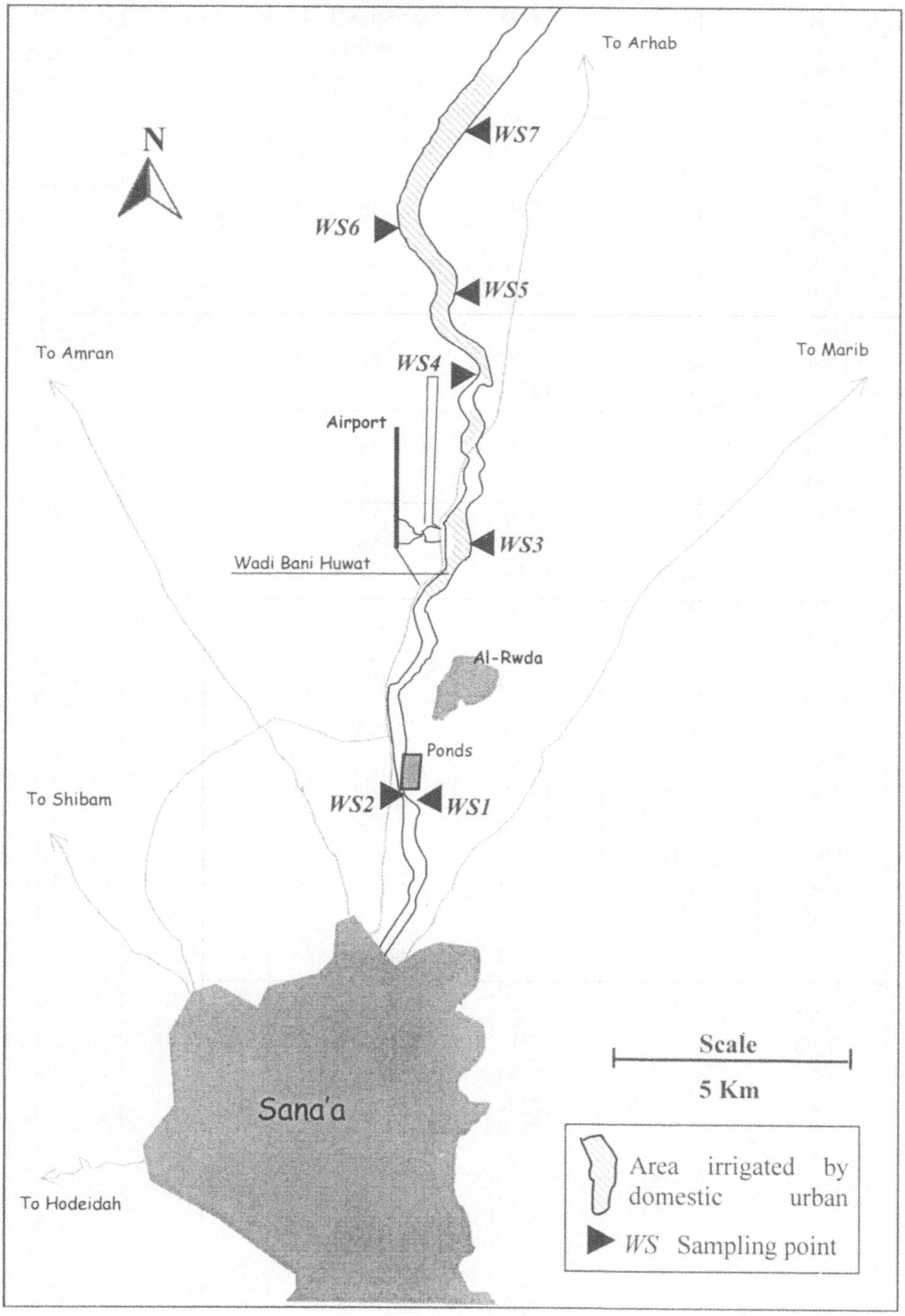

TableA2.1 Historic raw sewage analysis (results from NWASA Sana'a-branch laboratory)

| Date | BOD | COD | Suspended solids | Temp. °C | PH |
|---|---|---|---|---|---|
| 3/85 | 520 | - | - | - | - |
| 8/85 | 455 | - | - | - | - |
| 8/85 | 510 | - | - | - | - |
| 8/85 | 460 | - | - | - | - |
| 10/85 | 520 | 910 | 247 | - | - |
| 11/85 | 485 | - | - | - | - |
| 11/85 | 390 | - | - | - | - |
| 10/86 | 440 | - | - | 24 | 7.0 |
| 10/86 | 375 | - | 332 | 24 | 7.8 |
| 10/86 | 500 | - | 880 | 24 | 7.0 |
| 11/86 | 410 | 1800 | - | - | - |
| 7/87 | 470 | - | - | - | - |
| 3/88 | 475 | - | 414 | - | 7.7 |
| 3/88 | 280 | - | 102 | - | 7.6 |
| 3/88 | 535 | 1840 | 568 | - | 7.7 |
| 4/89 | 340 | - | 660 | - | 7.9 |
| 8/89 | 405 | 1666 | - | - | - |
| 8/90 | 405 | 1666 | 948 | - | - |
| 9/90 | 525 | 1532 | 1004 | - | - |
| 5/91 | 405 | - | 780 | - | 7.2 |
| 10/91 | 580 | - | - | 25 | 7.8 |
| 1/92 | 860 | 1736 | 647 | - | 7.9 |
| 2/92 | 710 | 1363 | - | - | 7.9 |
| 7/92 | 730 | 1441 | - | - | 7.4 |
| 10/92 | | 1650 | - | - | 7.6 |
| 1/93 | 700 | 1650 | - | - | 8.1 |
| 4/93 | 425 | 1640 | - | - | 8.1 |
| 5/93 | 657 | 1270 | - | - | 7.1 |
| 6/93 | 355 | - | - | - | - |
| 8/93 | 385 | - | - | - | 7.7 |
| 11/93 | 400 | 825 | - | - | 7.6 |
| 12/93 | 385 | 871 | - | - | 8.1 |
| 1/94 | 358 | 800 | - | - | 7.8 |
| 2/94 | 210 | 1206 | - | - | 8.1 |

Table A2.2 Evaluation of wastewater quality in Sana'a. (After Veenstra et. al, 1995)

| | $COD_{Total}$ | $COD_{Filtered}$ | $BOD_{Total}$ | $BOD_{Filtered}$ |
|---|---|---|---|---|
| Influent to ponds | 800 | 472 | 1607 | 576 |
| Effluent from ponds | 91 | 90 | 342 | 270 |

Table A2.3 Analytical results of raw sewage in Sana'a (Source, Howard Humphreys, 1995)

| Date | Time | BOD | COD | SS | $NH_4$ | Temp. °C | pH |
|---|---|---|---|---|---|---|---|
| 22/3/94 | 10.20 | - | 1206 | 724 | 124 | 23 | 7.7 |
| | 10.20 | 470 | 990 | 1148 | 124 | 23 | 7.7 |
| | 12.30 | 470 | 1205 | 1056 | - | 24 | 6.8 |
| | 12.30 | - | 1115 | 1796 | - | 24 | 6.8 |
| | 16.30 | 455 | 1351 | 880 | - | 24 | 6.9 |
| | 16.30 | 460 | 1102 | 812 | - | 24 | 6.9 |
| 30/3/94 | 7.20 | - | - | 476 | - | 21 | 7.6 |
| | 7.20 | 490 | - | 572 | - | 21 | 7.7 |
| | 12.30 | 550 | - | 804 | - | 23 | 6.9 |
| | 12.30 | - | - | 936 | - | 23 | 6.9 |
| | 16.30 | 540 | - | 804 | - | 23 | 6.9 |
| | 16.30 | 455 | - | 808 | - | 23 | 7.0 |
| 23/7/94 | 8.30 | 500 | 1208 | 364 | 102 | 25 | 7.9 |
| | 8.30 | 530 | 819 | 472 | 113 | 25 | 7.9 |
| | 12.00 | 540 | 1144 | 914 | 113 | 25 | 7.3 |
| | 12.00 | 530 | 1377 | 856 | 191 | 25 | 7.3 |
| | 17.00 | 500 | 1650 | 624 | 90 | 25 | 7.2 |
| | 17.00 | 500 | 1245 | 708 | 92 | 25 | 7.2 |
| 30/7/94 | 8.30 | 430 | 765 | 472 | 127 | 26 | 7.7 |
| | 8.30 | 450 | 778 | 392 | 137 | 26 | 7.7 |
| | 12.30 | 570 | 905 | 888 | 95 | 26 | 7.3 |
| | 12.30 | 550 | 994 | 756 | 99 | 26 | 7.2 |
| | 17.00 | 370* | 690* | 280* | 52* | 26 | 6.9 |
| | 17.00 | 350* | 516* | 212* | 57* | 26 | 6.9 |

*Diluted due to heavy rainfall.

Table A2.4 Location of sampling points

| Location | Code | Latitude | Longitude |
|---|---|---|---|
| Inlet of the existing oxidation ponds at Al-Rowda | WS1 | 1704195 | 416265 |
| Out let of the existing oxidation ponds at Al-Rowda | WS2 | 1704180 | 416045 |
| Near the village of Bait Al-Weshah | WS3 | 1709600 | 416830 |
| Next to the airport, at the site of the new treatment plant. | WS4 | 1712950 | 417085 |
| Near the village of Bait Al-Halaly. | WS5 | 1714440 | 416665 |
| Mid-way between the village of Bait Hanthal and Bait Sinhoob | WS6 | 1716070 | 415620 |
| Near the village of Bait Al-Baradi | WS7 | 1717905 | 416730 |

Table A2.4 Analytical results of wastewater samples

| Parameter | Sampling location | Date | | | | | |
|---|---|---|---|---|---|---|---|
| | | 18/5/96 | 28/5/96 | 15/7/96 | 31/7/96 | 23/11/96 | 8/1/97 |
| pH | WS1 | 7.48 | 7.80 | 7.76 | 7.61 | 7.68 | 7.90 |
| | WS2 | 7.83 | 7.96 | 7.71 | 7.84 | 7.46 | 8.08 |
| | WS3 | 8.27 | 7.96 | 7.70 | 7.52 | 7.34 | 7.47 |
| | WS4 | 7.96 | 7.80 | 7.74 | 7.21 | 7.32 | 7.59 |
| | WS5 | 7.99 | 7.84 | 7.87 | 7.41 | 7.42 | 7.70 |
| | WS6 | 7.87 | 7.96 | 7.74 | 7.59 | 7.62 | 7.71 |
| Temperature (°C) | WS1 | 24 | 25 | 25 | 24 | 21 | 20 |
| | WS2 | 23 | 24 | 22 | 21.5 | 20 | 18 |
| | WS3 | 24 | 25 | 25 | 22.5 | 19 | 19 |
| | WS4 | 24 | 25 | 24 | 22 | 19 | 18 |
| | WS5 | 25 | 24 | 24 | 21.5 | 18 | 18 |
| | WS6 | 25 | 25 | 24 | 21.8 | 18 | 18 |
| Electrical conductivity (μS/cm) | WS1 | 2365 | 2005 | 2320 | 2875 | 2530 | 2310 |
| | WS2 | 2287 | 2406 | 2461 | 2487 | 2145 | 2170 |
| | WS3 | 2279 | 2330 | 2380 | 2346 | 2060 | 2210 |
| | WS4 | 2141 | 2490 | 2488 | 2016 | 1990 | 2230 |
| | WS5 | 2200 | 2222 | 2477 | 1897 | 1900 | 2220 |
| | WS6 | 2160 | 2390 | 2345 | 2333 | 1820 | 2370 |
| Suspended solids (mg/l) | WS1 | 920 | 542 | 631 | 510 | 876 | 824 |
| | WS2 | 110 | 96 | 95 | 98 | 103 | 100 |
| | WS3 | 280 | 396 | 262 | 315 | 382 | 390 |
| | WS4 | 208 | 320 | 245 | 291 | 345 | 318 |
| | WS5 | 436 | 364 | 310 | 317 | 410 | 392 |
| | WS6 | 400 | 564 | 360 | 340 | 415 | 387 |
| Dissolved Oxygen (mg/l) | WS1 | 0 | 0.1 | 0 | 0 | 0 | 0.1 |
| | WS2 | 1.2 | 1.3 | 0.9 | 1.0 | 0.8 | 1.1 |
| | WS3 | 0.4 | 0.5 | 0.3 | 0.6 | 0.5 | 0.5 |
| | WS4 | 0.6 | 0.5 | 0.5 | 0.8 | 0.6 | 0.8 |
| | WS5 | 0.9 | 0.7 | 0.8 | 1.1 | 0.9 | 0.8 |
| | WS6 | 1.1 | 0.9 | 1.2 | 1.5 | 1.4 | 1.6 |
| $COD_{(filtered)}$ (mg/l) | WS1 | 1174 | 887 | 876 | 603 | 1379 | 813 |
| | WS2 | 751 | 419 | 445 | 427 | 475 | 390 |
| | WS3 | 455 | 730 | 844 | 956 | 725 | 1292 |
| | WS4 | 779 | 657 | 757 | 1140 | 785 | 1210 |
| | WS5 | 743 | 681 | 663 | 639 | 605 | 1164 |
| | WS6 | 564 | 668 | 651 | 842 | 485 | 1049 |

Continue Table A2.4

| Parameter | Location | 18/5/96 | 28/5/96 | 15/7/96 | 31/7/96 | 23/11/96 | 8/1/97 |
|---|---|---|---|---|---|---|---|
| Total Hardness ($mg/l$ as $CaCO_3$) | WS1 | 310 | 388 | 410 | 649.5 | 440 | 376.5 |
| | WS2 | 306 | 378.5 | 410 | 419 | 360 | 281 |
| | WS3 | 346 | 378 | 100 | 380 | 375 | 335 |
| | WS4 | 380 | 398 | 407 | 380 | 361 | 385 |
| | WS5 | 328 | 368 | 397 | 378 | 332 | 380 |
| | WS6 | 360 | 388 | 392.5 | 378 | 335 | 415 |
| Alkalinity ($mg/l$ as $CaCO_3$) | WS1 | - | 590 | 680.4 | 500 | 510 | 460 |
| | WS2 | - | 859.5 | 830.4 | 820 | 605 | 630 |
| | WS3 | - | 800 | 790.2 | 710 | 560.5 | 580 |
| | WS4 | - | 759.8 | 809.9 | 580.5 | 495.5 | 570 |
| | WS5 | - | 690 | 709.9 | 530.5 | 478 | 565 |
| | WS6 | - | 830 | 700 | 740 | 470 | 605 |
| Bicarbonate ($HCO_3$) (mg/l) | WS1 | - | 719.8 | 830 | 610 | 622.2 | 561.2 |
| | WS2 | - | 1048.6 | 1013 | 1000.4 | 738.1 | 768.6 |
| | WS3 | - | 976 | 964 | 866.2 | 683.8 | 707.6 |
| | WS4 | - | 927 | 988 | 708.2 | 604.5 | 695.4 |
| | WS5 | - | 841.8 | 866 | 647.2 | 583.2 | 689.3 |
| | WS6 | - | 1012.6 | 854 | 902.8 | 573.4 | 738.1 |
| Chloride (Cl) (mg/l) | WS1 | 268 | 223.4 | 246 | 557.9 | 410.6 | 336.3 |
| | WS2 | 251.3 | 269 | 268 | 290.3 | 226.6 | 254.9 |
| | WS3 | 245.7 | 269 | 264 | 264.1 | 215.9 | 223 |
| | WS4 | 281.4 | 311.5 | 281.4 | 223.4 | 215.9 | 219.5 |
| | WS5 | 276.8 | 237.2 | 339.5 | 219.1 | 276.1 | 215.9 |
| | WS6 | 276.8 | 272.6 | 290.3 | 259.1 | 230.1 | 226.6 |
| Sulfate ($SO_4$) (mg/l) | WS1 | - | 150.2 | 175 | 180 | 141.6 | 168 |
| | WS2 | - | 43.2 | 33 | 47 | 157.9 | 37.9 |
| | WS3 | - | 54.2 | 41 | 134.9 | 142.1 | 136.8 |
| | WS4 | - | 68.2 | 50 | 155 | 169 | 153.6 |
| | WS5 | - | 74.9 | 34 | 184.8 | 59.5 | 129.6 |
| | WS6 | - | 47 | 55 | 64.8 | 62.4 | 110.4 |
| Phosphate ($PO_4$) (mg/l) | WS1 | 14.7 | 22.2 | 24 | 26 | 24.7 | 26.9 |
| | WS2 | 47.2 | 22.5 | 10.5 | 7 | 30.4 | 6.7 |
| | WS3 | 20 | 22.5 | 23.7 | 40.6 | 34.9 | 35.5 |
| | WS4 | 24.3 | 19.7 | 21.7 | 37.5 | 48.5 | 30.4 |
| | WS5 | 46 | 22.2 | 21.7 | 55 | 56.7 | 39.6 |
| | WS6 | 39.3 | 23.1 | 19 | 58 | 48.2 | 57.1 |
| Nitrate ($NO_3$) (mg/l) | WS1 | - | 24.8 | 38.4 | 15.5 | 14.3 | 19.2 |
| | WS2 | - | 8.7 | 8.5 | 8.4 | 14.9 | 6.8 |
| | WS3 | - | 12.4 | 15.3 | 10.5 | 10.7 | 14.3 |
| | WS4 | - | 14.9 | 12 | 14.9 | 14.3 | 14.9 |
| | WS5 | - | 12.4 | 11 | 15.5 | 8.1 | 14.3 |
| | WS6 | - | 13 | 13 | 19.2 | 7.4 | 16.7 |
| Sulfur (S) (mg/l) | WS1 | - | - | - | - | 2.32 | 2.16 |
| | WS2 | - | - | - | - | 0.52 | 0.41 |
| | WS3 | - | - | - | - | 1.58 | 1.48 |
| | WS4 | - | - | - | - | 2.46 | 1.62 |
| | WS5 | - | - | - | - | 3.41 | 2.14 |
| | WS6 | - | - | - | - | 1.86 | 2.85 |

Continue Table A2.4

| Parameter | Location | 18/5/96 | 28/5/96 | 15/7/96 | 31/7/96 | 23/11/96 | 8/1/97 |
|---|---|---|---|---|---|---|---|
| Sodium (Na) (mg/l) | WS1 | 175.5 | 164 | 181.2 | 285.1 | 234.6 | 218.5 |
| | WS2 | 198.7 | 213.9 | 216.2 | 220.1 | 181.9 | 186.3 |
| | WS3 | 220.8 | 213.9 | 218.5 | 207.9 | 174.8 | 201.3 |
| | WS4 | 293.7 | 294.4 | 219.9 | 180.1 | 165.6 | 197.8 |
| | WS5 | 193.9 | 269.1 | 241.5 | 164.9 | 156.4 | 175 |
| | WS6 | 288.4 | 212.8 | 224 | 213.9 | 127.9 | 174.8 |
| Calcium (Ca) (mg/l) | WS1 | 72 | 108 | 108 | 204 | 128 | 104.6 |
| | WS2 | 78.4 | 112 | 116 | 120 | 108 | 76.4 |
| | WS3 | 94.4 | 108 | 112 | 116 | 112 | 92 |
| | WS4 | 96 | 112 | 112 | 116 | 94 | 106 |
| | WS5 | 83.2 | 104 | 108 | 108 | 90.4 | 106 |
| | WS6 | 80 | 112 | 108 | 108 | 86 | 116 |
| Magnesium (Mg) (mg/l) | WS1 | 31.7 | 28.8 | 34.2 | 34 | 29.3 | 28.1 |
| | WS2 | 26.8 | 24 | 29.3 | 29 | 22 | 22 |
| | WS3 | 26.8 | 26.4 | 29.3 | 22 | 23.2 | 25.6 |
| | WS4 | 34.2 | 28.8 | 31 | 22 | 30.7 | 29.3 |
| | WS5 | 29.3 | 26.4 | 31 | 26.4 | 25.9 | 28.1 |
| | WS6 | 39 | 26.4 | 29.9 | 26.4 | 29.3 | 30.5 |
| Potassium (K) (mg/l) | WS1 | 48.5 | 37.9 | 46.1 | 39 | 38.7 | 43 |
| | WS2 | 43.8 | 46.5 | 46.1 | 54 | 43 | 43 |
| | WS3 | 45.4 | 46.1 | 50 | 41.4 | 33.2 | 46.9 |
| | WS4 | 43.4 | 59.8 | 61 | 36.2 | 32.1 | 43 |
| | WS5 | 48.5 | 46.9 | 25 | 34.2 | 27.8 | 34.8 |
| | WS6 | 44.6 | 55.5 | 43 | 41.1 | 23.5 | 29.3 |
| Ammonium ($NH_4$) (mg/l) | WS1 | 202.5 | 129.6 | 159.8 | 87.7 | 91.8 | 113.4 |
| | WS2 | 151.2 | 74.9 | 144 | 161.8 | 113.4 | 117 |
| | WS3 | 169.6 | 158.8 | 131.9 | 138.6 | 109.8 | 100.8 |
| | WS4 | 106.4 | 154.8 | 164 | 109.1 | 113.4 | 86.4 |
| | WS5 | 128.3 | 129.6 | 131 | 111.6 | 110.2 | 106 |
| | WS6 | 130.9 | 162 | 130 | 150.3 | 98.1 | 109.8 |
| Boron (B) (mg/l) | WS1 | - | - | - | - | 1.7 | 1.3 |
| | WS2 | - | - | - | - | 1.4 | 0.9 |
| | WS3 | - | - | - | - | 2.3 | 1.2 |
| | WS4 | - | - | - | - | 1.5 | 1.6 |
| | WS5 | - | - | - | - | 0.9 | 1.1 |
| | WS6 | - | - | - | - | 0.8 | 0 |
| Calculated SAR | WS1 | 4.3 | 3.6 | 3.9 | 4.4 | 4.9 | 4.9 |
| | WS2 | 4.9 | 4.8 | 4.6 | 4.7 | 4.2 | 4.8 |
| | WS3 | 5.2 | 4.8 | 4.8 | 4.6 | 3.9 | 4.8 |
| | WS4 | 6.6 | 6.4 | 4.7 | 4.0 | 3.8 | 4.4 |
| | WS5 | 4.7 | 6.1 | 5.3 | 3.7 | 3.7 | 3.9 |
| | WS6 | 6.6 | 4.7 | 4.9 | 4.8 | 3.0 | 3.7 |
| Calculated $SAR_{(adj)}$ | WS1 | - | 10.0 | 11.0 | 12.4 | 12.9 | 10.6 |
| | WS2 | - | 13.7 | 13.4 | 13.4 | 11.3 | 12.9 |
| | WS3 | - | 13.6 | 13.6 | 13.0 | 10.6 | 12.9 |
| | WS4 | - | 17.9 | 13.6 | 10.9 | 9.3 | 11.9 |
| | WS5 | - | 16.6 | 14.6 | 9.9 | 8.7 | 10.6 |
| | WS6 | - | 13.4 | 13.7 | 13.4 | 6.7 | 10.3 |

Continue Table A2.4

| Parameter | Location | 23/11/96 | 25/11/96 | 7/12/96 | 9/1/97 |
|---|---|---|---|---|---|
| Total coliform (No. Colinies/100ml) | WS1 | 60E6 | 57E6 | 50E6 | 61E6 |
| | WS2 | 79E4 | 70E4 | 64E4 | 80E4 |
| | WS3 | 40E6 | 39E6 | 52E6 | 43E6 |
| | WS4 | 38E6 | 38E6 | 40E6 | 39E6 |
| | WS5 | 36E6 | 34E6 | 37E6 | 35E6 |
| | WS6 | 35E6 | 35E6 | 30E6 | 36E6 |
| | WS7 | 19E6 | 20E6 | 25E6 | 29E6 |
| Faecal coliform (No. Colinies/100ml) | WS1 | 28E6 | 29E6 | 31E6 | 30E6 |
| | WS2 | 50E4 | 38E4 | 45E4 | 56E4 |
| | WS3 | 22E6 | 20E6 | 20E6 | 24E6 |
| | WS4 | 19E6 | 18E6 | 21E6 | 21E6 |
| | WS5 | 16E6 | 13E6 | 19E6 | 15E6 |
| | WS6 | 14E6 | 12E6 | 15E6 | 14E6 |
| | WS7 | 10E6 | 9E6 | 11E6 | 14E6 |

# Acknowledgment

Thanks to Allah the exalted, the most merciful, for giving me the strength and persistence to keep going with this research even during the darkest moments.

I would like to express my thanks to my promoter, Prof. Alearts, for his guidance and support throughout this research. Special thanks to Prof. Walters who contributed greatly though useful and valuable discussions and suggestions. Many thanks to Dr. Braadbaart for his support, guidance and encouragement. Special thanks to Ir. Veenstra for his useful and critical advise, comments and discussions.

I would also like to thank all those involved in the Sana'a University Support (SUS) project, particularly Ir. Luiendijk for his assistance, understanding and continuous support. Many sincere thanks to Ir. Paul Post for his support and flexibility to overcome the obstacles during the fieldwork period of this research. I am grateful to Dr. Mohamed Al-Eryani and Dr. Abdullah Babaqi for their constructive comments and discussions. Special thanks are also extended to Prof. Savenije and Ir. Van der Gun for the useful discussions especially at the start of the research.

I deeply recognize the help and support of the staff of the National Water and Sanitation Authority (NWSA), Sana'a branch, particularly Mr. Ali Al-Yadumi, Mr. Abdu Jahafi and Mr. Mohamed Buragi, staff of the central laboratory, for their technical assistance.

I remain very grateful and gratified to my family, especially my mother, my wife and children for their support, patience, understanding and prayers throughout the period of this work.

# Biography

Mohamed Ibrahim Al-Hamdi was born in the small town of Thula, 40 km to the north of the capital Sana'a, Yemen, in September 1, 1966. In September 1983 he enrolled in the School of Engineering of George Washington University, Washington D.C., USA, where he obtained a B.Sc. in civil engineering in June 1988. Between September 1988 and October 1989 he joined the Yemeni Military Crops of Engineers for a one-year military service. During this time his duties included a variety of tasks connected to the design and supervision of civil works.

In September 1990, he joined the Faculty of Engineering of Sana'a University, where he worked as a teaching assistant covering laboratory work, general civil engineering subjects and water related subjects in particular. Within the framework of the Dutch funded Sana'a University Support (SUS) Project, he came to the Netherlands to study in the International Institute for Infrastructural, hydraulic and Environmental Engineering (IHE), Delft, in September 1992. He obtained a post graduate diploma in Sanitary Engineering in September 1993 and a M.Sc. with distinction in April 1994. Within the capacity building component of the SUS project he registered for Ph.D. on the sustainability of water resources in the Sana'a basin in August 1994. From June 1995 to September 1997 he conducted fieldwork in the study area besides carrying out some additional duties in the University. Since his return to Delft in 1997, he has been pursuing his Ph.D. research at IHE.

## Samenvatting

Sana'a, Jemen's hoofdstad, is gelegen in een "bekken" omgeven door heuvels op de centrale hoogvlakte van het land. Het bekken heeft een oppervlakte van 3,200 $km^2$. De bevolking in dit bekken is bijna geheel afhankelijk van grondwater voor zowel de stedelijke drinkwatervoorziening als voor irrigatie. De grondwaterwinning is intens sinds het begin van de jaren zeventig. In 1990 bedroeg de onttrekking ongeveer 180 $Mm^3$ waarvan wellicht ongeveer 83% voor irrigatie was geoormerkt. Het jaarlijkse aanvul-tekort van de grondwatervoorraad bedroeg ongeveer 120 $Mm^3$ in 1995, wat neerkomt op 300% van de totale natuurlijke aanvulling, hetgeen een jaarlijkse grondwaterpeilverlaging van 3-6m tot gevolg had.

Deze studie stelt zich niet tot doel een gedetailleerd actieplan te ontwerpen voor het aansturen van de vraag naar water (*demand management*) zodat de excessieve grondwateronttrekking onder controle kan worden gebracht. Eerder wordt hier gepoogd de belangrijkste incentief-systemen en -raamwerken te analyseren, die het gedrag en de beslissingen beinvloeden van de belangrijkste watergebruikers in het Sana'abekken: de stedelijke huishoudens en de boeren.

**Hoofdstuk 1**
Excessieve onttrekking van grondwater is een typisch "Common-pool-resource" (CPR)(gemeenschappelijke natuurlijke hulpbron) probleem, gezien eigendomsrechten ongedefinieerd zijn terwijl niemand de toegang tot de hulpbron ontzegd kan worden. In de competitie voor een CPR waarderen de gebruikers ervan toekomstige inkomsten die erdoor worden gegenereerd lager dan inkomsten die op korte termijn kunnen worden gerealiseerd. Onder dgl. omstandigheden geldt het recht van de snelste, en kan de hulpbron op korte termijn worden opgebruikt. Lokale gebruikers hebben weinig baat bij een gedrag dat bijdraagt tot duurzamer gebruik, als er niet een methode is die de hulpbron op een eerlijke, geordende (transparante) en efficiente wijze toewijst en verdeelt.

Om de competitie onder controle te houden en zo de waarde van de hulpbroneenheden te verhogen moeten eigendomsrechten beter worden gedefinieerd. Dit impliceert doorgaans dat twee soorten regels moeten worden bepaald en nageleefd: grensregels die de toegang tot de hulpbron afspreken, en gezagsregels die bepalen hoe de hulpbroneenheden worden verdeeld. Voor duurzaam gebruik moeten de laatste overeenstemmen met de de veilige productiecapaciteit. Eenmaal deze regels afgesproken, moeten ze worden nageleefd. Wanneer de winst groot is is de verleiding groot ze te breken, vooral wanneer niet gecontroleerd kan worden of anderen de regels wel naleven. Om dit probleem op te lossen moeten gebruikers een neutrale monitor accepteren en bereid zijn overtreders te straffen.

**Hoofdsuk 2** beschrijft de structuur en kostprijzen in de markt van huishoudelijk watervoorziening van Sana'a. De soort en het dienstniveau van de watervoorziening, en dus de prijs en het aandeel van water in de huishoudelijk budget, lijken bepaald door de woonplaats in de stad. Omdat de prijsverschillen tussen de verschillende typen watervoorziening groot zijn, zijn het arme huishoudens die afhangen van privé watertankers en gefilterd drinkwater van supermarkten die het meest betalen. Het aandeel van de waterdiensten (drinkwater, huishoudwater, en sanitatie) in een gemiddeld huishouden varieert van 1.5 tot 13.5%, afhankelijk van dienstniveau en inkomen. Door zijn lage kostrecuperatie en afnemende watervoorraad moet het publieke waterbedrijf zich beperken tot zijn huidige beperkte afdekking van de stad. Waterimport van andere regios brengt grote sociale, politieke, financiele en technische problemen met zich mee.

**Hoofdstuk 3** gaat in op de gevolgen van het lozen van ongezuiverd stedelijk afvalwater op de kwaliteit van het drinkwater in Sana'a, omdat kortsluitstromen worden vermoed via zowel diepere grondwaterlagen als via aanzuiging door onderdruk in het leidingnet. Meer dan de helft van de bevolking betrekt water van ongereguleerde waterverkopers die hun water aankopen van privé putten die her en der in de stad verspreid liggen. Dit is potentieel gevaarlijk omdat ongeveer 75% van de bevolking zijn afvalwater rechtstreeks in de ondergrond loost via latrines. Wij bevonden dat het patroon van grondwaterverontreiniging correleert met de bevolkingsdichtheid; grondwater onder de dichter bevolkte wijken bleek sterker vervuild dan dat onder minder dicht bevolkte wijken. Deze verontreiniging leidde een groot deel van de bevolking ertoe drinkwater te kopen van ongereguleerde kleine filterstations. De grote drukveranderingen in het openbare waterdistributienet van de wijk Nokom, dat niet continu water levert (net zoals de meeste andere delen van het netwerk), bleek negatieve druk te induceren en water aan te zuigen uit de bovenste lagen van de grond die gesatureerd is met afvalwater.

**Hoofdstuk 4** onderzoekt de legale en traditioneel-culturele status van grondwater in het Sana'abekken om beter inzicht te verkrijgen in de eigendomsstructuren, watergebruiksrechten en mechanismen om competitie-conflicten op te lossen. Door het gebrek aan een watercode in Jemen is grondwaterbeheer in feite overgelaten aan zijn gebruikers. De afwezigheid van een wettelijk raamwerk en werkbare institutionele mechanismen, gekoppeld aan een zwak overheidstoëzicht, hebben deze gebruikers *carte blanche* gegeven om het grondwater uit te putten. De grondwet verklaart alle natuurlijke hulpbronnen tot staatseigendom, maar de *shari'ah* daarentegen beschouwt grondwater een gemeenschapsbezit waarvan alle extern aangevoerde water niettemin (verhandelbaar) privé-eigendom is van de landeigenaar. Doorgaans wordt in het bekken grondwater aanzien als eigendom van de landeigenaar, maar in de praktijk wordt de toegang tot het water bepaald door het aandeel dat iemand heeft in de kosten van de uitbating van de put, ongeacht zijn landeigendom. De stam heeft een doorslaggevende rol in het dagelijkse leven en bepaalt in grote mate het raamwerk dat watertoewijzing en conflictbeheersing regelt. De *sheik* is dan ook een sleutelfiguur in alle gemeenschappelijke besluitvorming.

**Hoofdstuk 5** onderzoekt de agronomische aspecten van grondwatergebruik in de *qat*- en druivenkweek. De mechanisatie heeft grondwaterirrigatie snel uitgebreid sinds de jaren zeventig. De algemene gewoonte om gezamenlijk een put uit te baten heeft de toegang tot grondwater veel vergemakkelijkt, en heeft dus bijgedragen tot een redelijk eerlijke verdeling van de baten en een grotere productiviteit. De teelt van producten met hoge marktwaarde heeft de economische waarde van grondwater tevens laten stijgen. Nochtans is dit gegaan ten koste van de duurzaamheid. De boeren in het bekken gedragen zich rationeel maar blijken risico-avert in teeltkeuze, agronomie en irrigatiegedrag. Druiven kregen voorrang in irrigatieplanning, maar *qat* kreeg doorgaans meer water toebedeeld wat vaak leidde tot over-irrigatie. De beschikbaarheid van grondwater heeft het teeltpatroon sterk gewijzigd, met de omzetting van droge gebieden die traditioneel graan produceerden in *qat* en druivenproductie.

**Hoofdstuk 6** evalueert de kosten en baten die met grondwaterirrigatie gepaard gaan, voor zowel *qat* als druiventeelt. Het effect van hogere inputkosten wordt onderzocht. Buiten de kapitaalslasten voor land en initiële aanplant bleek het kostaandeel van water in de productiekosten van *qat* en druiven in de orde van 60-70% te liggen. De gemiddelde jaarlijkse teeltkosten lagen rond 2,000 en 1,500 US$/ha voor resp. *qat* en druiven. De prijsstijging van 67% van diesel in 1997 gaf aanleiding tot een stijging in het aandeel van de brandstofkost 28 naar 35%. Het kostaandeel van de brandstof in de globale productiekost bedraagt slechts 12-26%, dat van water-gerelateerde kosten 19-34%, en de O&M kosten 29-65%. De productiewaarde van land bedroeg ongeveer 8,000-13,000 US$/ha en die van het water 0.5-0.8 US$/$m^3$ voor qat, terwijl de waarden voor druiventeelt resp. 3,000-7,000 US$/ha en 0.4-0.8 US$/$m^3$ zijn. De kost/baat verhouding ligt voor *qat* en druiven op resp. 15-25% en 25-45%. Gegeven de grote winstmarge is het onwaarschijnlijk dat regelgeving gericht op de koststructuur grote slaagkans heeft. Waterbeheer dient derhalve te appeleren aan andere waarden, en in een meer participatieve en transparante wijze te geschieden.

**Hoofdstuk 7** analyseert de mogelijkheden tot hergebruik van Sana'a's huishoudelijk afvalwater in de landbouw. Het onderzoekt bestaande praktijken, afvalwaterhoeveelheden, en identificeert en waardeert zes hergebruiksscenarios op basis van vijf criteria. In 1997 bedroeg de hoeveelheid afvalwater ongeveer 5.2 $Mm^3$/jaar, of een afdekking van 20-25%. Afvalwater wordt door boeren opgepompt om voeder- en graangewassen te bevloeien op velden langs het afvalwaterkanaal. Momenteel verdwijnt ongeveer 75-80% van het stedelijk afvalwater in de ondergrond via latrines, hetgeen de grondwaterlagen artificiëel aanvult maar ze tevens verontreinigt. Boeren vinden het hergebruik financieel voordelig maar dit leidt nog niet tot minder oppompen van grondwater. Momenteel is de kwaliteit van het afvalwater onvoldoende om hergebruik toe te laten, maar uitbreiding van het rioleringsnet en de nakende ingebruikname van een grote, nieuwe afvalwaterzuiveringsinstallatie zou in principe meer hergebruik moeten toelaten. Analyse van de zes hergebruiksscenario's echter geeft aan dat grondwateraanvulling en her-oppompen voor het bevloeien van cash crops de beste optie is.

**Hoofdstuk 8** vat het incentief-systeem samen dat het niet-duurzame grondwatergebruik in het Sana'abekken aanstuurt. Factoren (dis-incentieven) die boeren beletten zelf-georganiseerde arrangementen op te zetten om het overpompen onder controle te brengen zijn: (1) de grootte van het bekken die de complexiteit van het probleem en de heterogeniteit van de stakeholders doet toenemen; (2) het gebrek aan juiste gegevens, met name tussen de gebruikers; (3) hogere waardering van korte-termijn baten; en (4) het onvermogen van het politieke bestel om de pompers faciliteiten en steun te bieden die kunnen toelaten dat het institutioneel systeem zich aanpast. De huishoudens aan de andere kant, blijken zich slechts te bekommeren om het probleem op micro-schaal (waarvandaan de volgende tankerlading te halen). De algemene informatievoorziening terzake blijft gebrekkig en onvolledig, en leidt tot de perceptie dat het grote-schaal probleem een probleem van de overheid is.